# MEASURING WHAT WE DO IN SCHOOLS

# MEASURING WHAT WE DO IN SCHOOLS

## How to Know If What We Are Doing Is Making a Difference

VICTORIA L. BERNHARDT

ASCD
Alexandria, Virginia USA

1703 N. Beauregard St. • Alexandria, VA 223111714 USA
Phone: 800-933-2723 or 703-578-9600 • Fax: 703-575-5400
Website: www.ascd.org • E-mail: member@ascd.org
Author guidelines: www.ascd.org/write

Deborah S. Delisle, *Executive Director;* Robert D. Clouse, *Managing Director, Digital Content & Publications;* Stefani Roth, *Publisher;* Genny Ostertag, *Director, Content Acquisitions;* Susan Hill, *Acquisitions Editor;* Julie Houtz, *Director, Book Editing & Production;* Joy Scott Ressler, *Editor;* Georgia Park, *Senior Graphic Designer;* Mike Kalyan, *Director, Production Services;* Andrea Hoffman, Senior Production Specialist; Keith Demmons, *Production Designer.*



PAPERBACK ISBN: 978-1-4166-2397-7 ASCD product #117021 n6/17
PDF E-BOOK ISBN: 978-1-4166-2398-4; see Books in Print for other formats.

Quantity discounts are available: e-mail programteam@ascd.org or call 800-933-2723, ext. 5773, or 703-575-5773. For desk copies, go to www.ascd.org/deskcopy.

**Library of Congress Cataloging-in-Publication Data**

Names: Bernhardt, Victoria L., 1952- author.
Title: Measuring what we do in schools : how to know if what we are doing is making a difference / Victoria L. Bernhardt.
Description: Alexandria, Virginia, USA : ASCD, [2017] | Includes bibliographical references and index.
Identifiers: LCCN 2017006919 (print) | LCCN 2017018390 (ebook) | ISBN 9781416623984 (ebook) | ISBN 9781416623977 (pbk.)
Subjects: LCSH: School improvement programs--Evaluation.
Classification: LCC LB2822.75 (ebook) | LCC LB2822.75 .B428 2017 (print) | DDC 371.2/07--dc23
LC record available at https://lccn.loc.gov/2017006919

25 24 23 22 21 20 19 18 17 1 2 3 4 5 6 7 8 9 10 11 12

# MEASURING WHAT WE DO IN SCHOOLS

# Preface

This is the twenty-first book I have published. Interestingly enough, a book about program evaluation was what I thought would be my first, and probably only, book. I have come full circle, back to the desire to share the tidbits of information that any teacher or administrator can use to improve what they do every day to help every student learn.

The Program Evaluation Tool described in this book has been used over and over to improve programs, processes, and schools. Through time and use, it has continuously improved. I am honored to share it through this ASCD forum.

Doing your own program evaluation or comprehensive data analysis work might be a challenge for many schools. Start where you are, by doing comprehensive data analysis with all your staff, or by analyzing one program; but continue building your analysis and knowledge about your school processes and the results they produce to become a true learning organization. The hard work will be worth the effort.

Thank you for your interest in improving what we do in schools. I encourage you to join me and get involved in *Measuring What We Do in Schools* because evaluation matters—to our teachers, administrators, other staff, parents, and, most of all, our students.

# Acknowledgments

I am a very lucky author because I have many people around the country who actually enjoy reviewing early drafts of my books, and who give willingly of their time and expertise. Listed below are those who provided thorough reviews that made the final product so much better. I appreciate what each and every one of you did to help make this book one of which I am proud. Special thanks go to Joy Rose and Susan Hill for their multiple reviews, edits, and suggestions; to Mary Younie for her graphic and formatting help; and to Wendy Heyd for data sets. A special thanks to the schools that gave me data to use in the examples. In concert with our agreements, I will not reveal your names or your real locations. We all appreciate the opportunity to learn from you.

I am pleased to as well acknowledge, with sincere thanks, the following reviewers: Jana Chang, Brad Geise, Sowmya Kumar, Kathy Miller, and Diane Yoshimura.

As always, I am appreciative of my husband, Jim Richmond, for providing his brand of support for my work. He does a lot of what I should be doing around the house so I can pursue these publications and the work to which I am so committed.

# 1

# Measuring What We Do in Schools

*How well does your school measure the work it does?* Most schools respond to this question by saying, "We are doing a really good job of analyzing our data." Unfortunately, many of these schools merely use every way under the sun to analyze their high-stakes test results. They have data walls, data binders, data teams, and send teachers to data retreats to help them analyze student performance on high-stakes tests.

Schools engage in these activities in good faith; however, many of these activities are counterproductive. The underlying assumption of working only on improving student achievement results on high-stakes tests is that only the students need to improve. The solution is often to offer tutoring, add after-school programs, or place students in special intervention classes. Many times the implementation of these strategies involves asking students to sit through their regular classes—although it is not expected that they will "get it"—and then to show up at the end of the day for lessons geared to their levels. These strategies result in spending more time repeating the same processes that produced the initial results.

What is inherently unsatisfactory with these "data use" trends is that they are too narrow. In effect, we might not be seeing the forest for the trees. Not looking at the school as a system can keep us from understanding the impact of our processes and the contributing causes of undesirable results, leaving us unable to eliminate their detrimental effect. When we listen to systems thinkers like Peter Senge, we hear them say that about 80 percent of what needs to change to get better results is *us*! (Senge, 2006.) Us, as in the *processes, practices, programs,* and *interventions* we are currently using that are producing these results. Evaluating our systems, processes, practices, and programs is the logical next step to seeing the forest *and* the trees.

## Learning Organizations, Continuous School Improvement, Systems Thinking, and Program Evaluation: How Do They Work Together?

*As learning organizations, "schools" are perfectly designed to get the results they are getting now. If "schools" want different results, they must measure and then change their processes to create the results they really want.*

—Adapted from W. Edwards Deming,
Continuous Improvement Guru, *Out of the Crisis*

Many schools would say they "do" continuous school improvement. Many of these same schools would describe their continuous improvement efforts in steps similar to these:

1. Study high-stakes test results by student groups;
2. Determine gaps among student groups;
3. Take last year's plan from the shelf and rewrite it to arrange interventions for the groups of students with the largest gaps;
4. Submit the plan to funding sources to secure funding for the interventions;
5. Print the plan, place it in a binder, and arrange it artistically on a shelf with the plans from previous years;
6. Hope the interventions will make a difference in student achievement; and
7. One year later, start again with step 1, above.

Unfortunately, as they plan for school improvement, many schools neither look at the big picture of where they are on measures beyond their most recent high-stakes test scores, nor do they study how they got their results to determine if new solutions might improve students' results. Paul Black and Dylan Wiliam (1998) sum up the conundrum quite eloquently:

> How can anyone be sure that a particular set of new inputs will produce better outputs if we don't at least study what happens inside?

These same schools would say they want to "do" a systemwide evaluation; they just don't have the resources, the time, or the know-how.

In this book, I hope to change school staff thinking on both continuous school improvement and evaluation, and give staff members tools to embrace systems thinking and become a true learning organization. I want program managers; teachers; and school, school district, and state and province administrators to come away from this book with an understanding of how to carry out comprehensive, systemwide evaluations of programs and processes in their learning organizations. I also want preservice teachers and administrators to use this book to get a firm grounding of the critical role program evaluation serves in school success, and learn how to implement meaningful evaluations that matter to schools. Specifically, this book will:

1. Present how continuous school improvement can guide evaluation and systems thinking in schools.
2. Show educators that evaluation work is logical and easy to do, giving them the confidence that they can—and should—be doing this work on a regular basis.
3. Shift thinking about evaluation from something that is done *to* educators to something *they need to do* on a continuous basis to make sure their work is making the intended difference.
4. Unpack a logical framework that teachers and administrators can use to evaluate programs, processes, and an entire school.
5. Expand teachers' and administrators' *data-informed decision-making* focus to include:
   — assessing what is working and what is not working for their students;

— determining which processes need to change to get better results; and
— using data, on an ongoing basis, to improve practices.

Before we tackle these goals, let's start by reviewing the definitions and intentions of the concepts *learning organizations, continuous school improvement, systems thinking,* and *program evaluation*. Then we must determine how the concepts work together, and why we must work with all these concepts in a comprehensive evaluation. As we progress, you will see what it will look like when we implement these concepts the way they are intended to work together.

## Learning Organizations

Learning organizations are, according to Senge, "Those organizations where people continually expand their capacity to create the results they truly desire, where new and expansive patterns of thinking are nurtured, where collective aspiration is set free, and where people are continually learning how to learn together."

Senge believes that only those organizations that adapt quickly and effectively can excel. In order to become a learning organization, two conditions must be present at all times. First, the organization must be purposefully designed to achieve the desired outcomes. In addition, the organization must develop the ability to recognize when its direction is moving away from the desired outcomes. The organization must have the capacity to get back on track. Organizations that are able to do this are exemplary. Learning organizations that can do this are continuously improving organizations.

Singapore American School is a true learning organization. Its mission is to "provide each student an exemplary American educational experience with an international perspective." The school's mission is to be "a world leader in education, cultivating exceptional thinkers, prepared for the future." To this end, the school embraces experiential learning, service learning, and personalized learning with high-impact, standards-based instructional practices. To do this with almost 4,000 preschool to grade 12 students from 50 nations, the staff has to continually learn and adjust.

As one administrator writes in his biography, "Singapore American School is a place that is nimble and courageous enough to support revolutionary change while upholding a tradition of proven excellence." (For more information on the Singapore American School, go to sas.edu.sg.)

## Continuous School Improvement

This is the ongoing effort to improve results, services, and processes. Continuously improving schools aspire to become learning organizations.

After a year's intensive training in continuous school improvement and comprehensive data analysis, leadership teams in Hawaii work to institutionalize continuous school improvement methods in everything they do. They start each new school year reviewing schoolwide data and expectations for the year and outcomes for the short-term and long-term. The entire staff analyze and understand why it is important to know schoolwide data. Together, they revisit the mission and vision of the school, and they discuss the implications in their grade levels and cross-grade-level teams. They establish assessment, instructional monitoring, and systemwide collaboration strategies to ensure that every student is learning, that every classroom is of highest quality, and that they have instructional coherence. The principal and instructional coaches meet regularly with teachers, grade-level teams, and the whole staff to ensure that the plans are being implemented with integrity and fidelity, and that every student is learning. At the end of the year, staff celebrate the gains made for every student, attributed to staff working smart and together.

## Systems Thinking

Systems thinking is the process of understanding how things influence one another within a whole. In learning organizations, systems consist of people, structures, programs, and processes that work together to help organizations get the results they want. Continuous improvement processes work hand in glove with systems thinking to ensure healthy learning organizations.

For example, designing a system of prevention—also known as response to intervention (RtI)—is a daunting task for most schools. One reason schools fail at this task is that they think of RtI as a program or a way to pull low-performing students out of classrooms to get help. RtI is a huge system; in fact, it must become the school system. To be successful, RtI requires systems thinking. Instead of considering RtI as a set of interventions for low-performing students, one must embrace it as the system of assessment and instruction in the school that helps every student grow in learning, every year. Using systems thinking, a schoolwide RtI team can review all classrooms' assessment results, and then determine interventions to ensure success for all students. This systems-thinking approach permits the school to adjust resources and materials across classrooms and grade levels, as needed, and determine if there are system weaknesses that need direct mediation.

## Program Evaluation

This is the systematic collection of information about the activities, characteristics, and results of programs to make judgments about the program, improve or further develop program effectiveness, inform decisions about future program development, and increase understanding (Patton, 2008). Evaluation of programs and processes is needed to improve an organization on an ongoing basis and to understand the impact of the parts and their interactions on the system. Continuous improvement and systems thinking use program evaluation to create the learning organization that can produce desired results.

When I work with schools in continuous school improvement training, they are asked to make a list of all processes in the school—instructional processes, such as differentiated instruction, grading, and student self-assessment; organizational processes, such as leadership teams, teacher collaboration teams, and instructional coaching; administrative processes, like discipline strategies, attendance programs, and retentions; and programs, which often include special education, after-school programs, gifted, and 9th grade academies. Schools usually come up with at least 60 programs and processes that operate in the school. Figure 1.1 presents a sampling of school processes and programs. Most teachers tell me that they do not know if any or all of their processes are making a difference—separately or together.

**FIGURE 1.1**

**School Processes and Programs**

| *Instructional Processes* | *Organizational Processes* | *Administrative Processes* | *Continuous School Improvement Processes* | *Programs* |
|---|---|---|---|---|
| • Academic conversations with students<br>• Classroom assignments (choices, projects)<br>• Classroom discussions (teacher talk, student-to-student talk, student-to-teacher talk)<br>• Collaborative groups<br>• Differentiated instruction<br>• Direct instruction<br>• Formative assessments<br>• Grading<br>• Homework<br>• Immersion<br>• Inquiry process<br>• Standards implementation<br>• Student reflection and self-assessment<br>• Summative assessments<br>• Technology integration<br>• Tutoring | • Data teams<br>• Data use<br>• Instructional coaching<br>• Leadership structure<br>• Mentoring<br>• Parent involvement<br>• Policies and procedures<br>• Problem solving<br>• Professional reflection<br>• Professional reflection discussions and support<br>• Professional Learning Communities<br>• Referral process<br>• Response to Intervention (RtI)<br>• Teacher collaboration<br>• Teacher observations<br>• Teacher renewal | • Attendance program<br>• Class sizes<br>• Data collection<br>• Discipline strategies<br>• Enrollment in different courses/ programs/ program offerings<br>• Leadership turnover<br>• Number of support personnel<br>• Policies and procedures<br>• Retentions<br>• Scheduling of classes<br>• Student groupings<br>• Teacher assignments<br>• Teacher certifications<br>• Teacher hiring<br>• Teacher and staff turnover | • Continuous school improvement planning<br>• Contributing cause analysis<br>• Data analysis and use<br>• Program evaluation<br>• Leadership<br>• Mission<br>• Partnerships<br>• Professional learning<br>• Self-assessment<br>• Vision | • Accelerated Reader/ Math<br>• After-school<br>• At-risk<br>• AVID<br>• Bilingual<br>• Dropout prevention<br>• English as a Second Language<br>• Gifted and Talented<br>• Honors<br>• International Baccalaureate<br>• Interventions<br>• Science Fairs<br>• Service Learning<br>• Special Education |

Next, staff are asked to analyze the processes and programs any way they would like. One often-used approach is to color code, as follows:

- *Green*. This process or program is important to our vision and everyone is implementing it as intended.
- *Yellow*. This process or program is important to our vision and not everyone is implementing it as intended.
- *Pink*. This process or program is optional or a duplication of efforts and needs to be tweaked to align to our vision.
- *Red*. This process or program is not important to our vision and should be eliminated.

After performing this analysis, participants often ask, *Does any school have all green-coded processes?* This is when we discuss what it would take to get all green-coded programs and processes in a school. It becomes obvious that a shared vision is required, along with clarity of what is most important to implement that shared vision. Next, we work together on clarifying the purpose, intended outcomes, and the way the program or process should be implemented.

We take one of the processes we know we want all staff to implement with integrity and fidelity and use the Program Evaluation Tool, which you will learn about in Chapter 3, and spell out the purpose, intended outcomes, whom the program or process is intended to serve, and how it should be implemented, monitored, and evaluated. Participants are delighted by what they learn about implementing with integrity and fidelity. When asked, "How many of these programs or processes should you spell out like this?" participants are unanimous as they call out, "*All* of them!" And then they say, "Wow, imagine the results we could get if the whole staff was implementing only the programs or processes that matter."

Program evaluation is at the core of continuous school improvement, systems thinking, and the learning organization. Consequently, if schools are not evaluating programs and processes, they are not using systems thinking or continuously improving. If schools are not using systems thinking and continuous improvement, they cannot create healthy learning organizations. Figure 1.2 shows the relationship between learning organizations, continuous school improvement, systems thinking, and program evaluation. Program evaluation is at the core. Program evaluation with systems thinking leads to continuous school improvement. Program evaluation, systems thinking, and continuous improvement are all necessary for schools to become true learning organizations.

**FIGURE 1.2**

**Relationship Between Learning Organizations, Continuous School Improvement, Systems Thinking, and Program Evaluation**

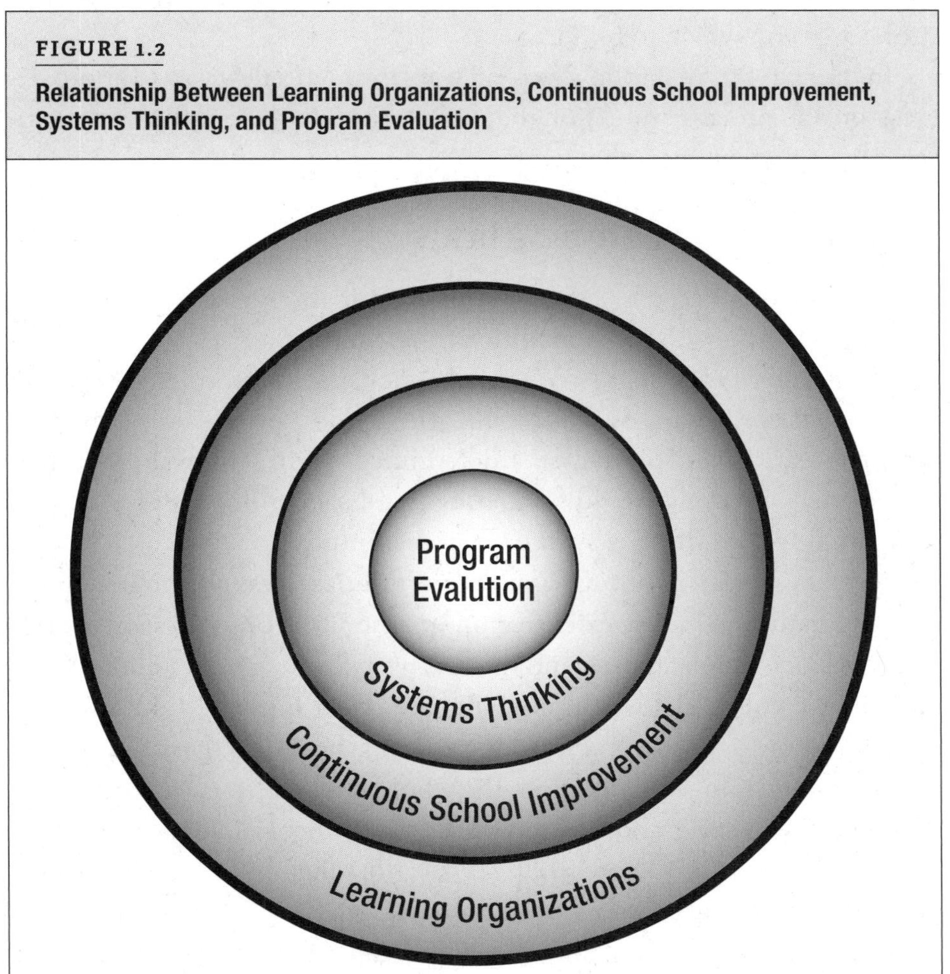


If we are a continuously improving school, we are using program evaluation and systems thinking, and we are aspiring to become a true learning organization. Continuous school improvement is the process of studying the impact of the parts (e.g., programs and processes) of a school and how the parts fit together to create the system, and then adjusting the parts so the intended outcomes are achieved. Program evaluation is used to acquire information from a variety of sources, study the impact of current programs and processes on overall organizational efficiency and effectiveness, and identify challenging areas that need to be strengthened, as well as strong areas on which to build.

In the chapters that follow, we will walk through a new way to approach program and process evaluation that pushes thinking, keeps the work manageable, and most important, works!

## How This Work Came About

For almost four decades, I have worked with thousands of schools to improve what they do for student learning. Starting as a program evaluator, at the end of each year I saw schools not using the wonderful reports I had worked hard to produce for them. The schools often said, "Next year is a new year. This is old news." I determined, then, that evaluation had to become something we do on an ongoing basis to improve for current students as well as those we will have in the future.

Several years later, through grant opportunities, I was able to design my own programs, including extensive program evaluations. These programs were considered outstanding and funders were flocking to my door to find out why we got results when programs they were funding did not. I told them, "I know everything about my program: how effective each part is, how the parts fit together to create the whole, the impact each part has on others and the whole, and the impact the system behind it has on the success of the program. I know what we have to adjust to maintain the purpose; and as we learn, I know when we can grow the purpose."

When you know a program so well, you can understand the impact the system that surrounds it has on the success of the program. My work naturally evolved to working with the whole school to ensure program success.

This work with effective programs led to applying these concepts of ongoing program evaluation to the larger organization: the whole school. The chapter that follows introduces the continuous school improvement framework that resulted from this work and has been used in numerous schools and countries. The framework describes how we look at the parts and the whole together to create the results we want to see in the school. The book builds from there in subsequent chapters to paint the complete picture of the tools and steps that are critical for continuous school improvement and evaluation.

## Structure of This Book

Chapter 2 presents the framework for continuous school improvement, which describes how program evaluation is required at all stages and is a good organizer for systems thinking and program evaluation work.

Chapter 3 explores the basic concepts surrounding program evaluation and outlines steps to prepare program evaluation using a special tool—the Program Evaluation Tool—that every school can and should use to design and evaluate individual programs and processes, even if there are limited resources for evaluation and no help from outside evaluators. The chapter also illustrates how to monitor program implementation. Program monitoring is a type of evaluation that allows staff to see the degree to which a program is being implemented. This ensures that the program is implemented with integrity and fidelity, and that the evaluation of the program is accurate and reflects what is actually occurring in classrooms, as opposed to what we think is occurring.

While measuring processes may seem difficult, the tools described in Chapter 4 make process evaluation easier to do and the effort worthwhile. The tools described in this chapter include the Process Measurement Planning Table and Flowcharting. The Process Measurement Planning Table helps us logically think through what we want a process to look like and how the processes can be measured. Flowcharting is a very useful tool for visualizing and communicating processes. When we add data to test the flowchart, we are able to assess if the process is doing what it is expected to do.

Chapter 5 looks at pulling it all together to evaluate the entire school. The chapter addresses what it will look like when all aspects of the school are assessed to understand if the school's Theory of Change is working, and how to determine what needs to change to get the desired results. Using the Theory of Change in evaluation work includes determining what data can be collected and using the data to implement only the processes and programs that a school knows will lead to its desired results. This chapter shows how the evaluation of different programs and processes using a logic model that is informed by the concepts and strategies outlined in previous chapters fit together into the continuous school improvement framework.

Chapter 6 describes how doing program and process evaluation work, in conjunction with schoolwide evaluation with systems thinking, can result in schools becoming true learning organizations. It also provides a few first steps to get you started.

As we progress through these chapters, you will see through templates and examples how the framework prompts us to ask big questions to guide our thinking, how the handful of tools we'll use helps us collect data critical to understanding our work and its impact, and how we pull it all together using a logic model that helps us zoom back out and piece together an accurate picture of our learning organization. The new clarity this process brings helps us better survey the landscape of our learning organizations and better lead our organizations in the right direction—toward improved student outcomes.

Finally, Chapter 7 tells the story of how one high school used all these concepts to change the way it does business to create better results. This school redesigned itself from a state-identified "focus school" residing in the bottom 20 percent of the state—in low student learning growth, proficiency rates on the state assessment, attendance, and graduation rates—to becoming a true learning organization, able to recognize when it is off track of its desired outcomes and what to do to get back on track.

# 2

# The Continuous School Improvement Framework: A Guide to Evaluation with Systems Thinking

*Knowing and not doing are equal to not knowing at all.*

Anonymous

Continuous school improvement is a logical way to bring ongoing improvement and evaluation together with systems thinking.

Continuous school improvement consists of five essential and logical questions that must be answered to create an organization that will make a difference for every student and every teacher, as well as increase student learning across the board each year. These questions are:

- Where are we now?
- How did we get to where we are now?
- Where do we want to be?
- How are we going to get to where we want to be?
- Is what we are doing making a difference?

Figure 2.1 shows these questions with the subquestions of the Continuous School Improvement Framework, all of which are described in the following sections.

**FIGURE 2.1**

**Continuous School Improvement Framework**

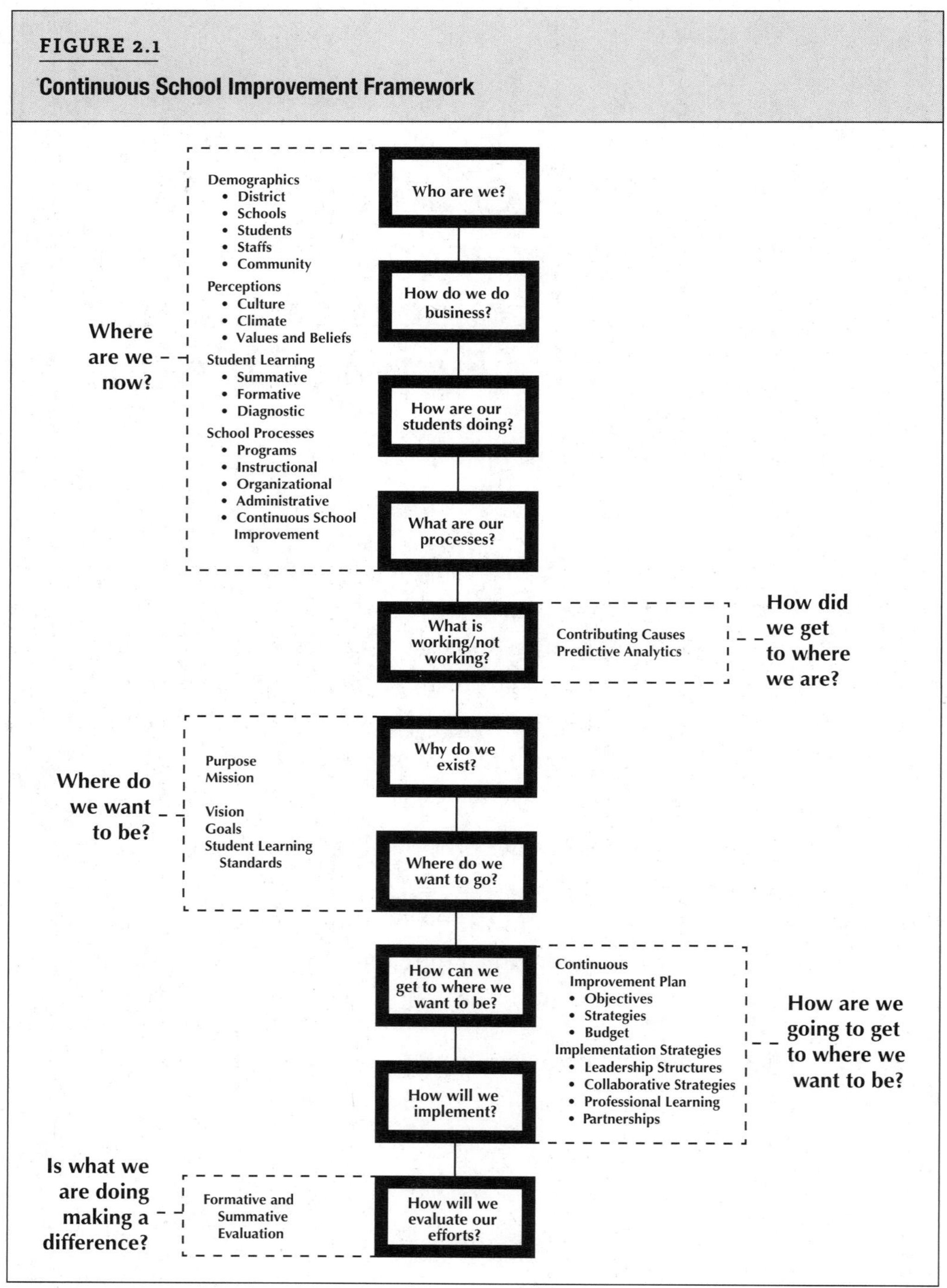

## Where Are We Now?

Continuous school improvement begins by looking honestly at where a school is right now with respect to four subquestions:

1. Who are we?
2. How do we do business?
3. How are our students doing?
4. What are our processes?

### Who Are We?

Answered with demographic data, this subquestion describes our students, staff, and community, and how these demographics have changed over time. Looking at the demographics over time is the best predictor of whom the learning organization will be serving in the future.

It is important to understand student and community demographics because it is the students whose needs we must meet. Student characteristics change as communities and demographics change. For instance, many communities across our country, and across other countries, are changing rapidly, going from having all-English or native-language-speaking students to a great percentage of non-English or native-language-speaking students, with influxes of immigrant families—some of whom have never experienced school. Demographic information can help teachers understand student demographic changes. Staff's professional learning needs to help them better meet the needs of their students, and can assist schools in placing appropriate teachers with students, hire strategically and effectively, and ensure a continuum of learning. All these data must be considered as staff members create a continuous school improvement plan that will make a difference for their students.

Gender, ethnicity/race, attendance, behavior, retentions, graduation rates, and program enrollment are demographic data that provide information about a school's students, as well as give the school insight into its leadership philosophies, and about how it moves students within the system. One surprising result of working with numerous schools' demographic data, on paper, is that I found I could predict, with about 96%

accuracy, how long a principal has been at any elementary school just by looking at demographic data and never having visited the school. My success rate is about 86% for secondary schools. What I saw changing the most were discipline referrals. Different leaders have unique philosophies of discipline that show up in the data. From there, I look at the relationship of discipline referrals to special education, and who is "allowed" to be gifted, assigned to Advanced Placement, or Honors, by gender and ethnicity/race. The philosophy of how we "treat" children shows up in demographics.

## How Do We Do Business?

This subquestion is answered through assessing a school's culture, climate, and organizational processes. Staff values and beliefs, most often assessed through questionnaires or determined during visioning processes, tell staff if team building or specific professional learning is necessary, what is possible to implement, and where a continuous school improvement plan must start.

One school I worked with was having a very hard time getting staff to change practices to better meet the needs of particular students who spoke English as a second language. A school improvement questionnaire was administered, which told us about some serious issues that needed to be addressed before change could occur. Over half the staff did not believe that all students could learn. They also did not believe that they knew how to work with the student population they had. Additionally, they reported they did not work well with one another. These questionnaire results gave us a different starting point for working with staff.

Student and parent questionnaires add other perspectives to the information generated from staff perceptions data. Students report what it takes for them to learn and how they are being taught and treated. Parents add information about what they need to help their children learn.

Another school I worked with blamed the students for lack of achievement. In questionnaire results, teachers stated that they felt that students and parents just didn't care about learning, and that all their efforts were for naught. Interestingly enough, the student questionnaire results indicated that students felt that the teachers did not care about or listen to them, and that they were not challenged by the work they did in class. Parent questionnaire results showed that parents did not feel welcome at the school, their children did not like to get up in the morning to go to school, and they wanted to know what they could do to help their children learn, because they saw their children failing in the classrooms and losing opportunities for productive futures. These questionnaire results changed the way teachers saw the students and their jobs. With strong leadership and professional learning support, they worked together to create a successful school that embraced all students. Perceptions help us see how the system is perceived from the inside out and the outside in.

## How Are Our Students Doing?

This subquestion requires a synthesis of student learning data, such as state assessment results, formative assessments, and teacher grades, in all subject areas, disaggregated by student groups, including gender, ethnicity/race, English learners, immigrants, migrants, students living in poverty, special education, grade level, following the same groups of students (cohorts) over time, and individual student learning growth. These data show whether schools are meeting the needs of all student groups and uncover strengths and areas for improvement. Student learning data should include all ways used by the school to determine that students are learning, which should be measuring what the learning organization wants the students to know and be able to do. Student learning data and this question help us see what the system is producing.

## What Are Our Processes?

This last subquestion in this section requires a complete accounting of all programs and processes operating throughout the school.

School processes include curriculum, instruction and assessment strategies, leadership, and collaboration. Programs include special education, extracurricular activities, gifted, advanced placement, and interventions. Programs and processes are the components of our organizations over which we have almost complete control, but we tend to measure these elements the least. A comprehensive analysis of school processes shows the extent to which the parts work effectively, independently, as well as together, to create the system that will achieve the desired results.

After completing an inventory of school programs and processes operating in their schools, as described in the example in Chapter 1, many staff see overlap. Together, they eliminate the overlapping programs and processes, and determine how to get all staff members working together to implement the programs and processes that will help them achieve their shared vision and their desired results.

## How Did We Get to Where We Are?—Contributing Causes

By understanding what is working and not working at school now, staff begin the analysis of how the school is getting its current results in all areas of student learning, so processes that are achieving the school's desired results are repeated, and strategies and structures that are not making a difference can be identified and eliminated. When staff review their schoolwide data, they start with demographics, perceptions, and school processes. Looking across these data's strengths and challenges, they can see why they are getting the results they are, and can almost predict student learning results. For example, when staff list a particular subgroup of students as a challenge, such as students living in poverty, they can see what this subgroup is saying about how they are learning and prefer to learn, and teachers can reflect on how they are teaching these students and understand the discrepancies. With this information, teachers determine what they have to do differently to get better results.

This second subquestion takes the analysis deeper to understand the contributing causes of the undesirable results, so schools can eliminate what is at the root of the undesirable results, and help inform the positive action to be taken. After conducting a contributing cause analysis, many teachers have come to understand that their processes have to change to better meet particular students' needs. While brainstorming 20 reasons they think this problem exists, staff's first 10 reasons often reflect their frustrations with not meeting the needs of the students and issues out of their control. Most often, teachers turn inward for the next 10 brainstormed reasons this problem exists. The last five are as close to the contributing causes as one can get, which are usually related to how the students are being taught—school processes. Figure 2.2 shows a list of 20 brainstormed reasons that a staff determined this problem exists in their school: Not Enough Students Are Growing in Learning.

**FIGURE 2.2**

**20 Brainstormed Hunches as to Why Not Enough Students Are Growing in Learning**

1. Too many students live in poverty.
2. There is a lack of parental support.
3. There is too much student mobility in our school.
4. The students aren't prepared for school.
5. Many of our students are not fluent in English.
6. Class sizes are too large.
7. Students don't do their homework.
8. Students do not like to read.
9. There is not enough help in the classroom.
10. Not all our curriculum is aligned to the standards.
11. We don't know what data are important.
12. We don't know how to use the data.
13. We don't know how to differentiate our instruction.
14. We are not teaching to the curriculum and standards.
15. Teachers are not teaching the content with fidelity.
16. Teachers don't know how to set up lessons to teach to student needs.
17. We need to know sooner what students know and don't know.
18. Our expectations are too low.
19. We need to collaborate to improve instruction.
20. Teachers need professional learning to work with students with backgrounds different from their own.

After staff brainstormed 20 reasons that the problem exists, they are asked to make a list of questions they need to answer with data to learn more about the problem. These questions generally come from the brainstormed reasons. They are also asked what data they need to gather to answer each question. An example follows as Figure 2.3.

FIGURE 2.3

**Questions and Data Needed to Know More About Why Not Enough Students Are Growing in Learning**

| *Questions* | *Data Needed* |
|---|---|
| 1. How many and who are the students who are not showing growth in learning each year? | 1. Student achievement results by grade level, subjects, student groups, classroom, and over time. |
| 2. How many students who do not show learning growth in English language arts also do not show learning growth in mathematics? | 2. Compare the answers above by subjects. |
| 3. What do the students who are not showing learning growth know and what do they not know? | 3. Student achievement standards and item analysis. |
| 4. What is the impact of our instruction? | 4. Student achievement by teacher processes. |
| 5. What do students, staff, and parents feel about the way students are learning? | 5. Student, teacher, and parent questionnaires. |
| 6. What do all our data tell us about student learning growth and what we need to do to improve? | 6. Study the data analysis results. |

The answers to these questions will inform staff about the problem and help them understand what they need to do to improve instruction for all students, which will result in more students showing growth in learning each year.

The first two major questions in the Continuous School Improvement Framework—*Where Are We Now?* and *How Did We Get to Where We Are?*—work together to create a comprehensive needs assessment.

## Where Do We Want to Be?—The Vision

A school defines its destination through its mission and vision, which should align with the district's mission and vision. Mission and vision are integral to everything a school does, and they have the power to influence programs and processes in large and small ways. Everything a school does should be aligned to its vision. Without a vision, a school's collective efforts have no target.

A school's mission must answer the subquestion, *Why do we exist?* The vision, created from the values and beliefs of members of the learning organization, and its mission, answer the subquestion, *Where do we want to go?* A clear vision is written in terms of the school processes to be implemented, including the curriculum, instruction, assessment, and the learning environment components required to carry out the purpose of the school. Many schools have a vision statement, but that is not enough to get the clarity required to get all staff members implementing the vision in the same way. Figure 2.4 shows the first pages of a four-part sample vision that asks teachers to identify components for curriculum, instruction, assessment, and the learning environment. This spelled-out vision describes what each component will look like when implemented, what has to happen at the school to get 100-percent implementation with integrity and fidelity, and the evidence that this component is being implemented. (A complete vision is shown in Figure 7.11.)

A continuously improving school has a vision that is clear and shared, and meets the needs of students. It is the vision that allows the continuously improving school to work cohesively as a system. It is through the creation and implementation of a vision that these systems-thinking schools can innovate and unleash greatness.

## How Are We Going to Get to Where We Want to Be?—The Plan

The answer to this question is key to unlocking how the vision will be implemented, and how undesirable results will be eliminated. An action plan—consisting of goals, objectives, strategies, activities, person(s) responsible, due dates, timelines, and resources required—needs to be developed to implement and achieve the vision and to eliminate the contributing causes of undesirable results. A continuous school improvement plan does not just provide action steps to close a gap; the continuous school

**FIGURE 2.4**

**Sample School Vision**

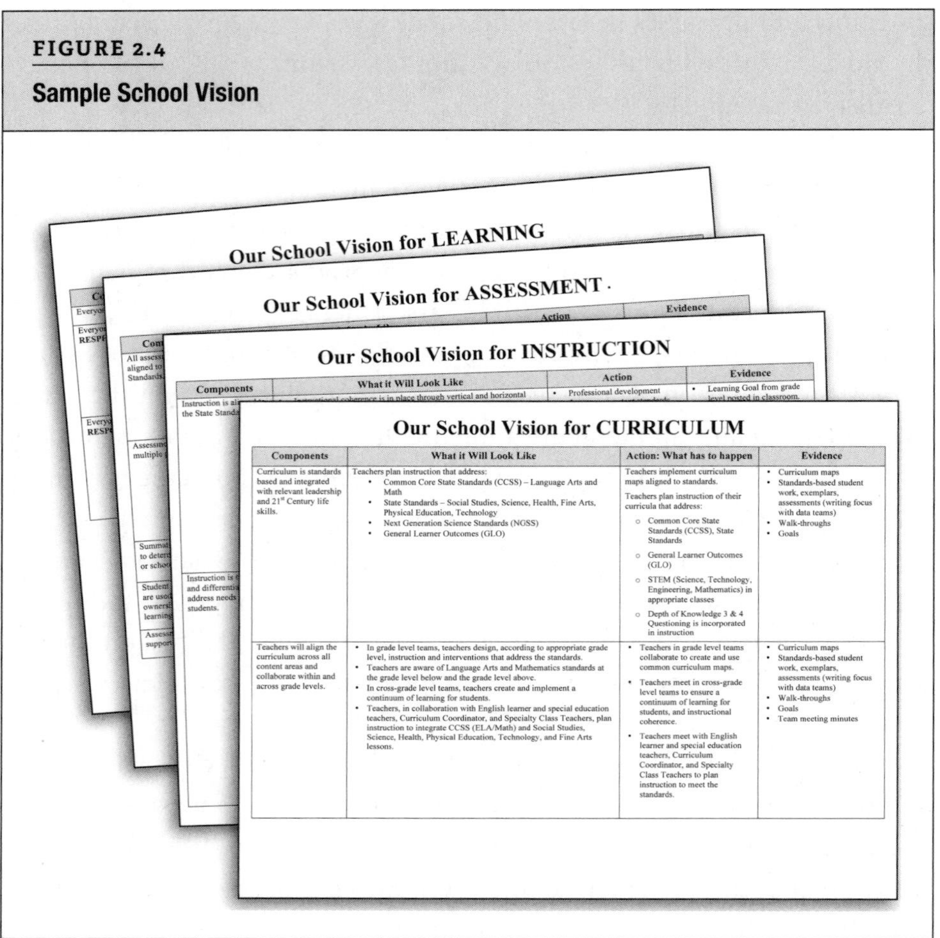

Our School Vision for LEARNING

Our School Vision for ASSESSMENT

Our School Vision for INSTRUCTION

Our School Vision for CURRICULUM

| Components | What it Will Look Like | Action: What has to happen | Evidence |
|---|---|---|---|
| Curriculum is standards based and integrated with relevant leadership and 21st Century life skills. | Teachers plan instruction that address:<br>• Common Core State Standards (CCSS) – Language Arts and Math<br>• State Standards – Social Studies, Science, Health, Fine Arts, Physical Education, Technology<br>• Next Generation Science Standards (NGSS)<br>• General Learner Outcomes (GLO) | Teachers implement curriculum maps aligned to standards.<br>Teachers plan instruction of their curricula that address:<br>○ Common Core State Standards (CCSS), State Standards<br>○ General Learner Outcomes (GLO)<br>○ STEM (Science, Technology, Engineering, Mathematics) in appropriate classes<br>○ Depth of Knowledge 3 & 4 Questioning is incorporated in instruction | • Curriculum maps<br>• Standards-based student work, exemplars, assessments (writing focus with data teams)<br>• Walk-throughs<br>• Goals |
| Teachers will align the curriculum across all content areas and collaborate within and across grade levels. | • In grade level teams, teachers design, according to appropriate grade level, instruction and interventions that address the standards.<br>• Teachers are aware of Language Arts and Mathematics standards at the grade level below and the grade level above.<br>• In cross-grade level teams, teachers create and implement a continuum of learning for students.<br>• Teachers, in collaboration with English learner and special education teachers, Curriculum Coordinator, and Specialty Class Teachers, plan instruction to integrate CCSS (ELA/Math) and Social Studies, Science, Health, Physical Education, Technology, and Fine Arts lessons. | • Teachers in grade level teams collaborate to create and use common curriculum maps.<br>• Teachers meet in cross-grade level teams to ensure a continuum of learning for students, and instructional coherence.<br>• Teachers meet with English learner and special education teachers, Curriculum Coordinator, and Specialty Class Teachers to plan instruction to meet the standards. | • Curriculum maps<br>• Standards-based student work, exemplars, assessments (writing focus with data teams)<br>• Walk-throughs<br>• Goals<br>• Team meeting minutes |

improvement plan creates the action to implement the shared vision, using the powerful information obtained from the schoolwide data analysis about the way the school is getting its current results.

## How Will We Implement?

The most powerful approaches to *implementing* a school vision and plan focus on four implementation strategies that provide the keys to implementing a healthy system:

- leadership,
- collaborative strategies,
- professional learning, and
- partnerships.

*Leadership* assists schools in creating decision-making structures to implement the vision. A leadership structure defines how teachers and administrators are going to implement the vision, and when they are going to collaborate to review progress toward the implementation of the vision. A leadership structure includes teachers—in a true learning organization, all teachers are leaders.

The leadership structure must look like the vision, or the vision will never be implemented. For example, a high school wanted to transform its subject-by-subject curriculum to interdisciplinary offerings. Staff could not figure out why it was not happening. Upon further inspection, we could see that while the new vision was very different from the previous vision, the leadership structure was the same. When the school adjusted its leadership structure to align better with the new vision, only then could the school effectively implement the vision.

*Collaborative strategies* are required in true learning organizations. Peter Senge says: "Collaboration is vital to sustain what we call profound or really deep change, because without it, organizations are just overwhelmed by the forces of the status quo."

Most schools have professional learning communities or data teams. It is critical that these structures help staff renew their practices on an ongoing basis. It is critical that teachers know how to review student data together and determine how to target instruction to eliminate learning needs. It is also crucial that these teams learn how to work with the big picture and not just one standard at a time. Just to meet and plan is not enough. Collaboration is necessary to implement the plan.

Leadership has to ensure that what is happening in collaborative teams and in grade levels is aligned to the vision, and that what is happening across grade levels is leading to instructional coherence. Instructional coherence is having a common framework that guides curriculum, instruction, assessment, and the learning environment, as well as horizontal and vertical articulation, which requires:

- Staff working conditions that support implementation of the common framework.
- Allocation of resources, such as funding, materials, time, and staff assignments, to advance the school's common framework.

Instructional coherence benefits students by promoting the integration of learning experiences and connecting those experiences over time. These connections can make learning experiences clearer, more meaningful, and more motivating. (Newman, Smith, Allensworth, & Bryk, 2001)

*Professional learning* helps teachers, administrators, and support staff engage in learning to continuously improve the manner in which they work—that is, how they make decisions; gather, analyze, and utilize data; plan, teach, and monitor achievement; collaborate with each other; evaluate personnel; and assess the impact of instruction and assessment on student learning. Professional learning should have one goal: to help everyone implement the vision.

*Partnership development* lays out the purposes of, approaches to, and planning for, educational partnerships with business and community groups, parents, and other educational professionals to increase student learning. When schools are clear about their vision and outcomes, others can more effectively contribute to that end. This is another place where leadership has to set the stage for effective partnerships to take place.

Partnerships can include student experiences in the workplace, collaboration with feeder schools to ensure a continuum of learning that makes sense to students, and field trips that align to classroom learning, to name a few.

## Is What We Are Doing Making a Difference?—Evaluation at All Stages

*Continuous improvement and evaluation* assists schools in understanding the interrelationships of the components of continuous school improvement and in improving their programs, processes, and products on an ongoing basis. *How will we evaluate our efforts?* refers to ongoing, formative evaluation to assess the effectiveness and alignment of all parts of the system to the vision so adjustments can be made to the system while it is operating. This question also refers to summative evaluation that reviews the results at the end of the year to see if the learning organization is getting its desired results. Both of these questions will be explored in detail throughout the rest of this book.

# 3

# Evaluating Programs

Schools have many reasons for not evaluating their processes and programs. Among the reasons schools give for not "doing" evaluations are:

- We don't evaluate. We *get* evaluated;
- It's hard, and we're not trained evaluators. It would take so much for us to learn, and we need to focus on our real jobs;
- Doing only a little bit of evaluation will not help, and we don't have the resources to do extensive evaluation;
- We do not know where to start;
- Program evaluation is valid only if it is done by people from outside the school;
- We've had bad experiences with program evaluation in the past;
- We don't have time to do it; and
- We don't think we will learn anything we don't already know.

We want to turn these thoughts around by engaging teachers and administrators in program evaluation that is useful, easy to perform, and that will energize their learning organizations to improve programs and processes for all students and make teaching more effective and efficient.

## Program Evaluation Defined

In Chapter 2, we reviewed the steps of continuous school improvement. Now we look again at continuous school improvement and think of it in terms of evaluation.

The evaluation we are referring to is called *program evaluation*, to distinguish evaluating programs and processes from evaluating people.

Many purposes exist for designing and carrying out program evaluation within a school or school district. These include, but are not limited to, concepts covered in this book:

- identifying what is working and what is not working *within* the school (program and process evaluation);
- assessing the degree to which programs, processes, and vision are being implemented as intended (program and process monitoring); and
- clarifying and assessing the effectiveness of the learning organization's overall vision (evaluating the learning organization).

## Evaluation Types

There are numerous types of program evaluation (please see References and Resources for further reading about the topic). The most common categories of program evaluation, and those we will target in this book, are formative, summative, program, and process.

### Formative Evaluation

This concerns how a program (i.e., a plan of action to accomplish a result) is *implemented*. It examines program activities, when program activities are implemented, how they are implemented, and how the program activities can be improved. Formative evaluation asks the following questions:

- Is the program being implemented as intended?
- What are the results based on what is being implemented?
- How can the program work better?

The ultimate purpose of formative evaluation is to understand how to improve the program or its processes—while the program is being implemented.

Schools use a form of formative evaluation when they assess student learning during an instructional unit to understand if students are internalizing the information. On the basis of the information, instruction is adjusted to meet the learning needs of students.

## Summative Evaluation

This addresses the results of a program or process. Summative evaluation takes place at the end of the program or process and assesses short-term as well as long-term outcomes. Summative evaluation examines if programs are achieving the desired results and how these results link to the processes used. Summative evaluation determines overall results and effectiveness.

High-stakes test scores are summative evaluations of how the students in a school are performing. These results tell us how the school did overall.

## Program Evaluation

As defined in Chapter 1, this is *the systematic collection of information about the activities, characteristics, and results of programs to make judgments about the program, improve or further develop program effectiveness, inform decisions about future program development, and increase understanding*. To these ends, program evaluation can be formative and summative.

## Process Evaluation

This is formative. It entails studying the strengths and challenges of the processes being used in the school. Process evaluation is also summative by determining if processes are producing the desired results.

For example, a school wanting to increase the number of students graduating from high school would put many structures in place. To understand what makes a difference with improving graduation rates, they would need to evaluate the impact of the strategies and activities, while determining their effect, separately and together, on the graduation rate.

In schools, we want to use a combination of formative and summative evaluations in our process and program evaluations. We use formative evaluation to continuously improve the implementation of our processes and programs. In the end, summative evaluation tells us if we got our desired results.

## Internal Versus External

Another common distinction made with regard to evaluation is whether it is internal or external.

### Internal Evaluations

These evaluations are performed by people within learning organizations—most often to plan for a program, to understand what is working and what is not working, and to determine how to improve a program or process. Internal evaluations can be formative as well as summative.

For example, data teams review formative data on an ongoing basis to understand what students know and do not know, how to better meet student needs, and if the strategies to improve student learning are working. In the end, they use related summative student learning data to know if students learned what was taught.

Principals also conduct internal evaluations as they walk through classrooms and schools looking for ways to improve student learning results. At the end of grading periods, they review the data to ensure that student learning results improved in every classroom.

### External Evaluation

This type of evaluation is conducted by people outside the learning organization, and provides an objective look at a program and its results. External evaluations are usually summative, although they can also be formative.

For example, a state assessment team might "evaluate" a school that is slated for school improvement. This usually consists of a team of colleagues from around the state visiting each classroom to understand how instruction is carried out. They compile reports about what they saw and what they think needs to be improved.

Many school personnel believe that evaluations are only "valid" if they are conducted by persons external to the learning organization. At the same time, there is a lot of disillusionment with external evaluations because it is often felt that external evaluators do not understand enough about the program being implemented to draw accurate conclusions about its implementation. When program staff feel that the external evaluations do not reflect a deep understanding of the program, they tend not to use the information from the external evaluators.

Internal evaluations can be just as valid as external evaluations, provided internal staff members have the skills to carry out the evaluation.

A summary of advantages and disadvantages of internal versus external evaluation is shown in Figure 3.1.

There is nothing worse than an external evaluator stating in a report that a teacher does not do something when, perhaps, it was just not observed during the brief classroom visit. While good external evaluators would conference before and after an observation and ask questions about what they are seeing, if they do not, their evaluation might be ignored. An internal evaluation is often the way to go. However, it must be done right. That means that it must be done only after leadership has won staff's trust, sufficient time has been dedicated to the evaluation, and it has been validated using an objective observation tool.

This book will enable teachers and administrators to conduct their own internal evaluations. If done in-house, a team, not an individual, should conduct the evaluation. A team of three to five people is ideal for most evaluations described here. An odd number of team members can help with reaching consensus. If individuals external to the organization do the evaluation, they should work closely with a school evaluation team to design and carry out an evaluation that will be accepted and used by staff.

Who should an internal evaluation team consist of? A principal or senior leader of an organization should lead the evaluation team as an impartial guide, with a teacher as co-chair. The leaders can see the changes that need to be made in the learning organization as the evaluation informs practice.

**FIGURE 3.1**

**Internal and External Evaluations: Advantages and Disadvantages**

**Internal Evaluation**

| *Advantages* | *Disadvantages* |
|---|---|
| Internal staff members are likely to understand the program or processes being evaluated. | The evaluation might not get done. |
| Internal staff members are likely to know and have some rapport with the people who need to be interviewed. | There could be some quality control issues, and a lack of expertise in program evaluation processes. |
| Staff are more likely to use the information if they or their colleagues conduct the evaluation. | It could be hard for internal staff to be completely objective. |
| Assigning someone within the organization can be less expensive than hiring someone from outside the organization. | It might be difficult for internal staff to evaluate programs with the big picture in mind. |

**External Evaluation**

| *Advantages* | *Disadvantages* |
|---|---|
| Evaluators external to the program can see the program with fresh eyes. | Those external to the program have less knowledge of the program and organization. |
| External evaluators might be more objective than internal staff. | It could take time for the external evaluators to develop trust among staff. |
| External evaluators probably have the experience and expertise in evaluation. | Staff might not trust outsiders and thus resist implementing their findings. |
| Time and money are allotted to external evaluation, so it gets done. | External evaluators might not be given enough time and information to understand programs with the big picture in mind. |

The leaders can also look across classrooms for instructional coherence. The teacher can help determine how these changes can be made with staff support, and how to reach the goal with staff. Depending on how big the learning organization is and what is being evaluated, representatives of each group implementing programs should be included, such as grade level and content area teachers. Individuals with the most knowledge of the program need to be included (e.g., in-house program trainers, instructional coaches). Certain program evaluations may require expertise in data analysis and questionnaire administration, as well as budgeting, if financial data are being included.

When staff do not initiate or participate in program evaluation work, they tend to think of evaluation as a negative process used for accountability and compliance. This creates an atmosphere in which evaluation is perceived as something that is done to them. If a school's own staff takes ownership of the process by doing the work of continuous school improvement themselves, evaluation becomes a part of the way work is done. Under these circumstances, schools are less likely to rely only on student learning data, and instead focus inward to find ways to improve learning for all students.

A few years ago, I was working with a school principal who was very proud of all the programs he was able to get funded in his school. He was also very proud of the fact that his K–5 school had healthy student achievement increases in some grade levels in certain subject areas during the previous year. The principal wanted me to find out which programs were making the most positive differences because he wanted to get all teachers using those programs and drop the programs that were not making a difference. I told him I would need some information about the programs being implemented, such as what was each program's intent, and to what degree any program was being implemented in each classroom. He showed signs of impatience with these questions because he thought he had hired me to just work with the results. I explained to him that the answers to these questions are powerful data and the only way we could know which programs or processes were making a difference. Much to his chagrin, he began listing the names and intentions of each program.

It didn't take long before he said, "Oh, my! If we were implementing the programs as intended, we would not need all of them, and it would be obvious which ones we should implement."

*If you can describe what a program will look like when implemented, you can implement it with integrity and fidelity, monitor its implementation, and evaluate its impact.* If you think that conducting a full-blown program evaluation is too difficult to do in-house, but you want to start planning for the evaluation of at least a few select programs, this chapter is for you. You might even discover you can do the full-blown evaluation.

After completing the continuous school improvement training, in which the Program Evaluation Tool is introduced, hundreds of leadership teams feel confident that they can perform program evaluation for many, if not all, of their programs and processes.

Many schools begin evaluations without doing the preparation work to make evaluation possible. The best preparation work is done as the program is being designed. Most well-intentioned evaluations end before they begin because evaluators realize the program results will be different depending upon how the program is being implemented by each individual. Without consistent implementation, evaluators of programs have no logical way to evaluate every combination and permutation.

We want program evaluation to tell us more than "it didn't work." We want program evaluation to remind staff of the intent of the program; help them understand what is working, what is not working, and why; remind them of how the program should be implemented compared to what is being implemented; and help them see the current results so they can continuously improve to make each program most effective.

The Program Evaluation Tool shown in this chapter helps staff create that program evaluation structure while designing the program, which prepares staff to:

- implement the program with integrity and fidelity;
- monitor program implementation;
- determine the support needed to implement with integrity and fidelity;

- determine results; and
- align program implementation to enhance the overall system.

The Program Evaluation Tool gets all staff on the same page with the intent of the program, how it should be implemented, and the expected results. If the program is already in operation, you can still evaluate the current status of program implementation, but you may have to redirect staff toward a modified implementation and a renewed or corrected understanding of the intent of the program.

## The Program Evaluation Tool

This does everything we want program evaluation to do, such as:

- remind staff of the intent of the program and whom it is intended to serve;
- help staff understand what is working, what is not working, and why;
- remind staff of how the program should be implemented compared to what is being implemented; and
- help staff see the current results so they can continuously improve to make each program most effective.

As shown in Figure 3.2, the Program Evaluation Tool is also designed to tie in the comprehensive needs assessment work that the school does as a part of its continuous school improvement efforts.

The full version of the Program Evaluation Tool consists of six major headers and twelve guiding questions. Figure 3.3 presents a sample District Mathematics Program that guides our understanding of the questions and headers that follow (and that are included in Figures 3.2 and 3.3).

### Needs Assessment

***What are the data telling you about the need for the program or process?***

What do the needs assessment data tell you about the current results and the need for this program or process? To describe the current situation, use multiple measures of data, specifically those described in the

**FIGURE 3.2**

**Program Evaluation Tool**

| ***Needs Assessment*** | ***Purpose*** | | ***Participants*** | ***Implementation*** | | ***Results*** | |
|---|---|---|---|---|---|---|---|
| *What are the data telling about the need for the program or process?* | *What is the purpose of the program or process?* | *What are the intended outcomes?* | *Whom is the program/process intended to serve?* | *How should the program be implemented to ensure attainment of intended outcomes?* | *How is implementation being monitored?* | *How will results be measured?* | *What are the results?* |
| | | | ***Who is being served? Who is not being served?*** | | ***How should implementation be monitored?***<br>***To what degree is the program being implemented with integrity and fidelity?*** | | |
| ***Implications for the Continuous School Improvement Plan:*** | | | | | | | |

**FIGURE 3.3**

**District Mathematics Program**

| **Needs Assessment** | **Purpose** | | **Participants** | **Implementation** | | **Results** | |
|---|---|---|---|---|---|---|---|
| *What are the data telling about the need for the program or process?* | *What is the purpose of the program or process?* | *What are the intended outcomes?* | *Whom is the program/process intended to serve?* | *How should the program be implemented to ensure attainment of intended outcomes?* | *How is implementation being monitored?* | *How will results be measured?* | *What are the results?* |
| • 50% of our students did not show learning growth on the State Assessment in Mathematics.<br>• Only 50% of our students were proficient on the State Assessment in Mathematics.<br>• Not every teacher is teaching the agreed-upon mathematics curriculum.<br>• The high school to which our middle school feeds has a 12% dropout rate. Of the students going on to post-secondary education, 68% require remediation in mathematics. | What is the purpose of the program or process?<br>• Increase the percentage of students proficient in mathematics on the State Assessment.<br>• Ensure a K–12 continuum of learning in mathematics.<br>• Provide ongoing professional development to improve mathematics instruction in every teacher's classroom.<br>• Support teachers in their classrooms to ensure that every student grows in mathematics every year. | When the mathematics program is implemented, the following will result:<br>• Every student will grow in mathematics knowledge every year.<br>• Teachers will feel confident in mathematics instruction in their classrooms.<br>• Teachers will know what students know and do not know so they can target instruction accordingly.<br>• Mathematics proficiency percentages will increase each year. | The mathematics program is intended to serve every student and every teacher in our school.<br>***Who is being served? Who is not being served?***<br>Students proficient and growing in mathematics learning are being served by the Mathematics Program.<br>Teachers improving in mathematics instruction are being served by the mathematics program.<br>Students not proficient or growing each year in mathematics are not being served. | • Every teacher in the district will attend two days of training in mathematics instruction at the beginning of the school year, and one day every month.<br>• Teachers will attend training, by grade level, so the training will focus on the grade-level standards, curriculum, and processes across all schools.<br>• School grade-level teams will also be trained in how to work together and support each other in implementing the mathematics program. | Program implementation is being monitored through instructional coaching and professional development.<br>***How should implementation be monitored?***<br>The mathematics program provides a self-assessment monitoring tool that also helps teachers implement with integrity and fidelity. Coaches will use the same tool during observations and coaching. Teachers and coaches will discuss discrepancies and how to improve instruction. We will start using the monitoring tool in the second semester. | Our school mathematics questionnaire will be administered four times a year to monitor teacher and student perceptions of teaching and learning mathematics.<br>Our school Mathematics Assessment will be given throughout the year to assess what students know and do not know, and if they are growing in learning.<br>The State Assessment in Mathematics will be used to determine proficiency and student learning growth, in the spring of each year. | Early results are showing student engagement to be improving.<br>Progress monitoring results indicate that almost every student is showing learning growth in mathematics —not at the same rate, but there is growth.<br>The system of instructional improvement is being embraced. Teachers are committed to improving their instruction.<br>There is evidence that grade-level teams are being effective and that the ongoing professional learning is being well received. |

| **Needs Assessment** | **Purpose** | | **Participants** | **Implementation** | | **Results** | |
|---|---|---|---|---|---|---|---|
| *What are the data telling about the need for the program or process?* | *What is the purpose of the program or process?* | *What are the intended outcomes?* | *Whom is the program/process intended to serve?* | *How should the program be implemented to ensure attainment of intended outcomes?* | *How is implementation being monitored?* | *How will results be measured?* | *What are the results?* |
| • Teachers reported on our school questionnaire that they do not feel confident teaching mathematics.<br>• Students responded on our school questionnaire that they feel they are not challenged by the work they are asked to do in school. | • Provide an assessment tool so teachers know what students know and do not know.<br>• Increase schoolwide student engagement and achievement.<br>• Guarantee that students in our feeder pattern will not require post-secondary mathematics remediation. | • A continuum of learning, K–12, will be evident—what students learn in one year will prepare them for the next year, and ultimately college.<br>• No student will need post-secondary mathematics remediation.<br>• Every student will be engaged in learning mathematics. | Teachers not improving in mathematics instruction are not being served. | • Every teacher will assess where students are at the beginning of the year, for a baseline, and every week, for Progress Monitoring, using our school Mathematics Assessment.<br>• School grade-level teams will review the assessments each week and help one another adjust instruction to better meet the needs of every student. | ***To what degree is the program being implemented with integrity and fidelity?***<br>Because the mathematics program has not been officially monitored, we do not know the degree of implementation. | The mathematics program self-assessment monitoring tool will be used, in part, to determine how teachers are implementing the program.<br>Instructional coaches will also monitor implementation.<br>Teachers and instructional coaches will conference about their two responses. | |

***Implications for the Continuous School Improvement Plan:***

- We need to pilot the monitoring tool this year so we can start the new school year with the tool in place in every classroom.
- We need to rethink how we do professional development next year. It needs to be ongoing, but not a repeat of what the teachers got this year. At the same time, we need to provide professional development for teachers new to our system.

Continuous School Improvement Framework: demographics, perceptions, school processes, and student learning. By looking at all types of data, you are able to see the big picture as opposed to just one element, such as student learning results. The answer to this question should help staff see the compelling reason for having and implementing the program. It assists with creating outcomes. The needs assessment results also provide a baseline for measuring improvement later on.

The first column of Figure 3.3 is the summary needs assessment for the District Mathematics Program. If this school staff looked only at student learning results, they would see that just half of their students were proficient, and half showed learning growth on the state assessment in mathematics. Those are pretty compelling reasons to implement a mathematics program. With this information, we would want the program to build the skills of all students in mathematics. However, when we look at all types of data gathered, we see that teachers have not agreed on what mathematics curriculum to use. Not only that, they do not feel confident teaching mathematics. Further, students do not feel they are challenged by the work they are asked to do in school, and the high school to which the school feeds has a 12-percent dropout rate. Of the students going on to post-secondary education from this high school, 68 percent require remediation in mathematics. With this comprehensive look, the sense of urgency to improve is very compelling, and we cannot offer a program that builds only student skills. The program must have a leadership component to get all teachers on the same page with a mathematics curriculum, build teachers' skills and confidence, and ensure a continuum of learning that transcends this school through college, as well. The starting and ending points are very different when we use multiple measures of data, as opposed to only one measure—student learning results.

## Purpose

***What is the purpose of the program or process?***

Describe the intent of the program so *anyone* can understand it, and so *everyone* on staff can understand it—in the same way. Make sure the program purpose is "large" enough to eliminate the undesirable results uncovered in the needs assessment.

Figure 3.3 includes the purpose of the District Mathematics Program, which is to:

- Increase the percentage of students proficient in mathematics on the State Assessment.
- Ensure a K–12 continuum of learning in mathematics.
- Provide ongoing professional learning to improve mathematics instruction in every teacher's classroom.
- Support teachers in their classrooms to ensure that every student grows in mathematics every year.
- Provide an assessment tool so teachers know what students know and do not know.
- Increase schoolwide student engagement and achievement.
- Guarantee that students in our feeder pattern will not require post-secondary mathematics remediation.

Even though we are not seeing districtwide results in this example, one can see that the issue of mathematics proficiency is larger than this one school can deal with, alone. However, one school can do its part to help with that next level of schooling. This is a good purpose. It is multi-faceted and reflects the needs assessment.

Figure 3.4 presents three purpose statements. The first—*to improve test results on the State Assessment in Mathematics*—is too basic. Some schools make a statement like this when setting out to evaluate a program. It just isn't big enough. Many teachers prefer a purpose statement like the one in the column titled Too Unclear—*to offer options that help students learn to their ability*. They say they don't want to dictate to others how to do their jobs, and that some students are only capable of learning so much. Well, that purpose statement is pretty telling about why that school is not getting great results. Chances are its results are very uneven across grade levels. The purpose in the third column is just right because it reflects the needs assessment and details what has to happen in the program and school to eliminate the undesirable results that appear in the needs assessment.

**FIGURE 3.4**

**Purpose Statements**

| *Too Basic* | *Too Unclear* | *Just Right* |
|---|---|---|
| The purpose of the District Mathematics Program is to improve test results on the State Assessment in Mathematics. | The purpose of the District Mathematics Program is to offer options that help students learn to their ability. | The purpose of the District Mathematics Program is to:<br>• Increase the percentage of students proficient in mathematics on the State Assessment.<br>• Ensure a K–12 continuum of learning in mathematics.<br>• Provide ongoing professional learning to improve mathematics instruction in every teacher's classroom.<br>• Support teachers in their classrooms to ensure that every student grows in mathematics, every year.<br>• Provide an assessment tool so teachers know what students know and do not know.<br>• Increase schoolwide student engagement and achievement.<br>• Guarantee that students in our feeder pattern will not require post-secondary mathematics remediation. |

Purpose is extremely important as a school works to get all staff members on the same page. It must be large enough to uncover the reason for doing the hard work and add that sense of urgency to do it honestly and right. If teachers know the purpose going into the program, they will think as big about the program as the purpose leads them, and have a chance to achieve all of those purposes. If they are not thinking big going into the program, they will only think about the purpose in their minds—most often, getting better student achievement scores. Imagine all of the unintended positive outcomes they could achieve by thinking bigger.

***What are the intended outcomes?***

Intended outcomes are what we want to happen as a result of the program. Outcomes are what we are trying to achieve. Make sure they are broad enough to cover the items mentioned in the needs assessment and are reflective of the purpose. Figure 3.5 shows three intended outcomes. The first one—*Test results on the State Assessment in Mathematics will improve*—is far too basic. Of course, we want to see state mathematics assessment results improve, but that is not all. The second column outcome—*Students will achieve to their abilities because of how teachers are teaching*—is too unclear. Not only do we not know what this person means about student abilities, we don't know how teachers are teaching. This analysis starts getting into methods for implementation—which are unclear as well. The intended outcomes in column three are just right because they reflect the needs assessment and the purpose of the District Mathematics Program, as follows:

*When the District Mathematics Program is implemented:*

- *Every student will grow in mathematics knowledge every year;*
- *Teachers will feel confident in mathematics instruction in their classrooms;*
- *Teachers will know what students know and do not know so they can target their instruction accordingly;*
- *Mathematics proficiency percentages will increase each year;*
- *A continuum of learning, K–12, will be evident; what students learn in one year will prepare them for the next year and, ultimately, college;*
- *No student will need mathematics remediation in college; and*
- *Every student will be engaged in learning mathematics.*

As with purpose, when outcomes are spelled out and big enough, staff can begin implementation with the complete picture. There is a better chance of achieving intended outcomes when they are spelled out.

## Participants

***Whom is the program intended to serve?***

Describe whom the program is intended to serve. Make sure the intended recipients, purpose, the needs assessment, and implementation are aligned. For example, many schools state that the purpose of their Response to Intervention (RtI) system is help *all* students achieve. Their needs assessment also shows that *all* students need to achieve learning growth. The way RtI is implemented is to provide interventions to the lowest scoring students, which leads to the next two questions.

**FIGURE 3.5**

**Intended Outcomes**

| *Too Basic* | *Too Unclear* | *Just Right* |
|---|---|---|
| Test results on the State Assessment in Mathematics will improve. | Students will achieve to their abilities because of how teachers are teaching. | When the District Mathematics Program is implemented:<br>• Every student will grow in mathematics knowledge every year.<br>• Teachers will feel confident in mathematics instruction in their classrooms.<br>• Teachers will know what students know and do not know so they can target their instruction accordingly.<br>• Mathematics proficiency percentages will increase each year.<br>• A continuum of learning, K–12, will be evident; what students learn in one year will prepare them for the next year and, ultimately, college.<br>• No student will need mathematics remediation in college.<br>• Every student will be engaged in learning mathematics. |

***Who is being served? Who is not being served?***

Describe whom the program is serving and who is not being served by the program. This is powerful information to assist with the improvement of the program. Given the example above, which shows that the purpose of RtI is to help *all* students achieve, because of the way this RtI program is designed and implemented, it cannot help *all* students grow. Therefore, the lowest-performing students are being served, and all the others are not being served.

In our District Mathematics Program example (Figure 3.3), staff indicated that the program is intended to serve every student and every teacher in the school. Students proficient and growing in mathematics learning are being served by the mathematics program. Teachers improving in mathematics instruction are being served by the mathematics program. Students not proficient or not growing each year in mathematics are not being served. Teachers not improving in mathematics instruction are not being served.

## Implementation

***How should the program be implemented with integrity and fidelity to ensure attainment of intended outcomes?***

Describe how the program will be implemented. Describe the steps so that *anyone* can understand how to implement with integrity and fidelity. A flowchart can be used to visualize the steps in the implementation process. (Flowcharting is described in Chapter 4.) Perhaps even a video of the program being implemented with integrity and fidelity can be used to clarify for all staff members what they are expected to do.

***How is implementation being monitored?***

Describe how implementation of the program is currently monitored. Very often, monitoring is left out of the implementation equation. Monitoring implementation in the classroom is the only way to know how, or if, the program is being implemented with integrity and fidelity. Our example shows that program implementation is currently being monitored through instructional coaching and professional learning, which is embedded within the school day.

***How should implementation be monitored?***

Describe how the program should be monitored to ensure implementation with integrity and fidelity. Very few schools will say they are actually monitoring the way implementation *should* be monitored. Our example school indicated its program should be monitored as follows: *The mathematics program provides a self-assessment monitoring tool that also helps teachers implement with integrity and fidelity. Coaches will use the same tool during observations and coaching. Teachers and coaches will discuss discrepancies and how to improve instruction. We will start using the monitoring tool next year.*

***To what degree is the program being implemented with integrity and fidelity?***

Provide the evidence that the program is being implemented with integrity and fidelity. (Note any discrepancies in the Implications or Next Steps section, at the bottom of the form.) This is valuable and necessary information for an evaluator to have when assessing the impact of the program. If the program is implemented with integrity and fidelity throughout the school, it is easy to understand how the school gets its results. If the program is not implemented with integrity and fidelity throughout the school, there will be unintended consequences. A program administrator would have to know under which circumstances the results came out as intended. Our example indicated a truth: *Because the mathematics program has not been officially monitored, we do not know the degree of implementation.*

## Results

***How will results be measured?***

Designate what processes or tools will be used to analyze results. The district's mathematics program example stated the following:

- *Our school mathematics questionnaire will be administered four times a year to monitor teacher and student perceptions of teaching and learning mathematics.*
- *Our school mathematics assessment will be given throughout the year to assess what students know and do not know, and if they are growing in learning.*
- *The state assessment in mathematics will be used to determine proficiency and student learning growth, annually.*
- *The mathematics program self-assessment monitoring tool will be used, in part, to determine how teachers are implementing the program.*
- *Instructional coaches will also monitor implementation using the self-assessment monitoring tool.*
- *Teachers and instructional coaches will conference about the two responses to the monitoring tool.*

***What are the results?***

Report the results of the program here. Use the *Intended Outcomes* and the *How Will Results Be Measured* columns to align results to outcomes and how you plan to measure results (shown in Figure 3.6 for our example mathematics program). While the three columns are not identical, you can see how the results are describing the other two columns, even though the program is just beginning.

**FIGURE 3.6**

**Aligning Intended Outcomes and How Results Will Be Measured**

| *Intended outcomes* | *How will results be measured?* | *What are the results?* |
|---|---|---|
| • Every student will grow in mathematics knowledge every year.<br>• Teachers will feel confident in mathematics instruction in their classrooms.<br>• Teachers will know what students know and do not know so they can target their instruction accordingly.<br>• Mathematics proficiency percentages will increase each year.<br>• A continuum of learning, K–12, will be evident; what students learn in one year will prepare them for the next year and, ultimately, college.<br>• No student will need mathematics remediation in college.<br>• Every student will be engaged in learning mathematics. | • Our school mathematics questionnaire will be administered four times a year to monitor teacher and student perceptions of teaching and learning mathematics.<br>• Our school Mathematics Assessment will be given throughout the year to assess what students know and do not know, and if they are growing in learning.<br>• The State Assessment in Mathematics will be used to determine proficiency and student learning growth, annually.<br>• The mathematics program self-assessment monitoring tool will be used, in part, to determine how teachers are implementing the program.<br>• Instructional coaches will also monitor implementation in the classroom and across grade levels.<br>• The District will follow student cohorts through high school graduation and into college. | • Early results show improvement in student engagement.<br>• Progress monitoring results indicate that almost every student is growing in mathematics —not at the same rate, but there is growth.<br>• The system of instructional improvement is being embraced. Teachers are committed to improving their instruction.<br>• There is evidence that grade-level teams are working and that the ongoing professional learning is being well received. |

## Implications for the Continuous Improvement Plan

Describe big-picture next steps that will be undertaken as a result of the work above.

As a staff completes the Program Evaluation Tool, they will encounter issues that need to be addressed to truly implement and evaluate a program with integrity and fidelity. This section is for keeping a running record of these next steps, which often need to be addressed in the continuous school improvement plan. Even if the next steps that emerge will not be a part of the continuous school improvement plan, this is a handy section for jotting down next steps. Rename the section if desired.

Our example District Mathematics Program team used this section to jot down the next steps that emerged while completing the Program Evaluation Tool. They stated:

- *We need to pilot the monitoring tool this year so we can start the new school year with the tool in place in every classroom.*
- *We need to rethink how we do professional learning next year. It needs to be ongoing, but not a repeat of what the teachers got this year. At the same time, we need to provide professional learning for teachers new to our system.*

## A Program or Process Is Only as Good as What Gets Implemented

*If you are not monitoring and measuring program implementation, the program probably does not exist.*

A couple of years ago, I was asked to perform a quick analysis of an arts-integration program that was being implemented by some teachers in a middle school. The principal thought that all I needed to do was to analyze the student achievement results for the classrooms in which teachers were implementing the program and compare those results with the classrooms in which teachers were not implementing the program. I did that analysis, which showed no differences. The principal was shocked. We discussed what teachers did when they were fully implementing the program. In learning more about the program, I found that some teachers had gone through ongoing intensive training and had been implementing the program for nine years. Others were just getting started. Teachers obviously were in various stages of implementation, with anywhere from nine months to nine years of training.

We set up a self-assessment tool to ask the teachers to what degree they thought they were implementing the concepts of the program. We did the data analysis by the teachers' assessment of their degree of implementation. Still no differences were found; in fact, results were very inconsistent. We had the trainer of the program look at the analysis and the results. The trainer did not agree with the teachers' self-assessments; therefore, the program trainer conducted her own assessment of the degree to which each teacher was implementing the program. We used the trainer's assessment to make the next analysis. With this analysis, we saw apparent differences.

Here is what we learned during this evaluation, the concepts of which have been validated many times since:

- To really know if a program is making a difference, we have to determine the degree to which the program is being implemented in every classroom, and the degree to which it is being implemented as intended.
- Teacher self-assessment is not an accurate depiction of the degree of implementation. After all, if teachers knew what 100-percent implementation looked like, they would probably be doing it. In fact, teachers often tell us the reason they are not implementing a program is because they do not know what it would look like, sound like, or feel like if they were implementing the program fully.
- One can use a teacher self-assessment instrument to support the implementation of a program. Outside observations and demonstration lessons are also needed to help teachers change their practices and perceptions of 100-percent implementation.
- The optimum assessor of degree of implementation is the program trainer—the person who knows the program best.
- The next best assessor of the degree of implementation is someone who knows the program and what it is supposed to look like when implemented. Very often the job of assessor is left to

the principal, who may have little or no training in assessing program implementation.

## Assessing Program Implementation

How can we assess program implementation, especially if we do not have access to the program trainer? I recommend a classroom assessment tool constructed by those most knowledgeable of the program, along with strong accountability for implementation. If an official assessment tool does not already exist, I recommend those knowledgeable of the program construct one as follows:

1. *Describe what the program should look like when it is fully implemented.* What will the teachers be doing? What will students be doing? Describe the curriculum, what the instructional practices will look like, how students will be assessed, and what the environment will feel like. How will resources be used? What will happen if the intended results are not being achieved?
2. *List the key elements that should be observable in the classroom.* For example, planning and preparation, classroom environment, instruction, professional responsibilities, and what students will be doing when the process is implemented with integrity and fidelity.
3. *Either construct a rubric that shows how teachers can begin implementing the vision and eventually evolve to 100-percent implementation, or establish a scale that indicates the degree to which these elements are seen in the classroom.* Figure 3.7 shows an assessment tool with a five-point scale to remind teachers of what it would take to implement a mathematics program—from not implementing the concept at all to implementing all the time. (Instructional coaches use this same tool to validate the teachers' self-assessments. The coaches' observations and collaboration with the teachers help ensure the implementation of the program in every classroom.)
4. *Agree on how you will use the instrument.* Some schools have teachers conduct their self-assessments, and compare the self-assessment with outside observers' observations—usually an instructional

**FIGURE 3.7**

**Teacher Assessment Tool for Measuring Implementation of District Mathematics Program**

To what degree are you implementing the District Mathematics Program in your classroom? Circle the number that represents the degree of implementation right now *(1 = not at all; 2 = some of the time; 3 = about half of the time; 4 = almost all of the time; 5 = all of the time).* Add comments about what would help you fully implement the District Mathematics Program, or notes about successes you have had that you would like to share with others.

| | ***1 2 3 4 5*** | ***Comments/Notes*** |
|---|---|---|
| I feel confident teaching mathematics. | 1 2 3 4 5 | |
| My instruction is standards-based. | 1 2 3 4 5 | |
| I create lesson plans that reflect the agreed-upon mathematics program. | 1 2 3 4 5 | |
| I assess what students know and do not know on an ongoing basis, using multiple strategies. | 1 2 3 4 5 | |
| I use activities that challenge all my students. | 1 2 3 4 5 | |
| I maximize instructional time to help every student meet content standards. | 1 2 3 4 5 | |
| I use multiple strategies to engage students in their study of mathematics. | 1 2 3 4 5 | |
| I ensure that students know and use the vocabulary of mathematics. | 1 2 3 4 5 | |
| I provide opportunities for students to demonstrate knowledge of mathematics content in a variety of ways. | 1 2 3 4 5 | |
| My instruction incorporates the concepts of rigor: conceptual understanding, procedural skills and fluency, and application. | 1 2 3 4 5 | |
| I showcase student work on a regular basis. | 1 2 3 4 5 | |

*continued*

**FIGURE 3.7** *(continued)*

**Teacher Assessment Tool for Measuring Implementation of District Mathematics Program**

| | ***1 2 3 4 5*** | ***Comments/Notes*** |
|---|---|---|
| I collaborate with colleagues to improve the way time and energy are spent in the classroom. | 1 2 3 4 5 | |
| I collaborate with colleagues to develop a K–12 continuum of learning in mathematics. | 1 2 3 4 5 | |
| I seek coaching or feedback from a colleague to continuously improve my practice. | 1 2 3 4 5 | |
| Students feel confident that they can "do" math. | 1 2 3 4 5 | |
| Students are engaged in the math work. | 1 2 3 4 5 | |

coach or another teacher. The self-assessment is useful for helping teachers understand what they are supposed to be implementing, and to help others understand what teachers understand and do not understand about implementing the program. The results are analyzed by grade level to determine professional learning or coaching support needs. The grade level information is shared in leadership team meetings to determine if the entire staff is having difficulty implementing similar concepts, or if other grade-level or subject-area teachers can help with specific concepts.

A rubric can also be used to support the implementation of programs and processes. This "homemade" rubric gives staff members the opportunity to get involved in the creation of the program and to set individual and staff-wide goals for improvement. The rubric, the Teacher Self-Assessment Tool, and staff collaboration can all serve as great collaborative approaches to evaluating program or process implementation. In addition, process flowcharts, described in the next chapter, show a picture of how the program is intended to be implemented. One can also create a process flowchart of how the program is being implemented now, which is helpful for improving program and/process implementation. Bringing the data to bear on the flowchart could show that perhaps the program is not getting the results it could be getting because it is not being implemented as intended. The data will also show how the variations of implementation can appear in the results. Let's take a closer look at flowcharts in Chapter 4.

# 4

# Measuring Hard-to-Measure Processes

Teachers and administrators of Dry Creek Elementary School, down the road from where I live in Northern California, asked me to come to their school to help analyze their data to find out what they needed to do to get all students reading on grade level by grade 3. I looked at how they had been analyzing their data. They did a great job with the data they had, but something was not quite right—something was missing.

I asked the teachers, "How do you teach reading?" They looked at me with almost disappointed eyes and said, "We thought you were going to help us analyze our data." I assured them that I was there to help them with their data. I also assured them that the answer to my question was important data to have. How they taught reading and what they did when students were not reading were major indicators of how they got their results.

Here is what they told me about how reading is taught in their school. In first grade, students are given a reading exam. The student scores are rank-ordered. The students scoring in the bottom 20 percent of the class were placed in special one-on-one instruction. The rest of the students were taught in the regular classroom. I asked why 20 percent were selected for the individual instruction. They said, "That is how you do the program." We followed the data and saw that those 20 percent from grades 1 and 2 did well on the third-grade high-stakes exam. We looked back at the rank-ordered list and added a grade-level-standard cut point. It was shocking to see that almost 65 percent of the first graders were not reading "on grade level," but were passed along because they did not happen to be in the bottom 20 percent. The same was true in grade 2, except the percentage of students not reading on grade level increased each year. The teachers could see that moving the bottom 20 percent into one-on-one instruction was not helping them reach their ultimate goal of getting all students reading on grade level by grade 3. The teachers created a flowchart of what they needed to do when students were proficient, and what they needed to do when students did not meet the proficiency cutoff, starting with their early-in-the-year diagnostic tests. By creating the flowchart, teachers agreed on appropriate instruction for students when they did or did not meet particular standards. One month later, teachers reassessed the students and saw that they were making progress. The teachers met to adjust their practices to get even better results. Three months down the road, they assessed again and happily saw that their improved practices were helping all students achieve.

## It's Time to Measure Our Processes

School processes define what learning organizations, and those who work in them, are doing to help students learn: what they teach, how they teach, how they assess students, and what they do when results fall short of expectations. School processes include programs, curriculum, instruction and assessment strategies, interventions, and all other classroom practices that teachers use to help students learn. School processes describe what teachers do when students are proficient, and what teachers do when students are not proficient.

To understand the student achievement results schools are getting, administrators and teachers must document and measure the processes being implemented. That information, aligned to the results, will help them understand how they are getting their results. Clarifying how they are achieving their successes will help teachers and administrators understand which processes are working and should continue. Understanding the processes that are not getting desirable results will inform teachers and administrators of which processes should be changed or eliminated. To truly know if a reading program is successful, a school has to know how reading is taught for every student, in every classroom, throughout the school.

Measuring processes is one of the most important things we can do to improve K–12 education. Processes are the only things over which we have extensive control in education. While some people may think measuring processes is difficult, the work can be accomplished, and it can be manageable. Measuring the processes used in instruction is a task we all have to work on to understand process impact and determine how to improve teaching and learning. If the implementation of specific processes is not being measured or monitored, those processes are probably not being fully implemented. Districts and schools must devote time to the management and measurement of school processes so that the successful ones can be shared and implemented school and districtwide, and unsuccessful practices can be redesigned or eliminated.

## How to Measure Processes

There are many ways to measure processes. Qualitative and quantitative measures can be applied. Qualitative school process measures might include program/course flow, focus groups, interviews, and questionnaire results. Quantifiable school process measurement can include classroom observations, program enrollments, and student achievement results. It is a common assumption that only one measure is sufficient, but multiple measures always provide more comprehensive data and information and paint a clearer, fuller picture. We need to think logically about what we need to know about a process and match these thoughts and questions to the measurement.

The Process Measurement Planning Table (Figure 4.1), designed to help you think about and plan for the continuous improvement of your successful current processes, illustrates how logical it can be to measure the hard-to-measure processes. Figure 4.2 presents some of the more common examples of processes that present challenges when it comes time to measure their success and effectiveness: curriculum, instruction, assessment strategies, staff collaboration, and learning environment. The Process Measurement Planning Table describes what we want the processes to look like when implemented with integrity and fidelity—which gives insight into the purposes of the processes, and examples of the many different ways in which each process can be measured. For example, if measuring curriculum, we probably want to align the curriculum to content standards and grade level expectations; we would need to find evidence that there is a continuum of learning that makes sense for the students, and that it is being implemented in every classroom. We can measure these through curriculum mapping, process flowcharting, classroom/teacher observations, student learning results, questionnaires, and assessing the implementation of the vision.

**FIGURE 4.1**

**Process Measurement Planning Table**

| *What process do you want to measure?* | *What do you want the process to look like?* | *How can this process be measured?* |
|---|---|---|
| | | |

Let's zoom in a little closer on the last hard-to-measure process—integrity and fidelity. What would that data look like? It could consist of multiple pieces of qualitative data that can be ultimately verified with quantitative data, such as student learning results. The qualitative data could include an implementation monitoring tool (as shown in Figure 3.7). That tool could be used as a self-assessment, coupled with observations by an instructional coach. The Program Evaluation Tool (similar to those in Figures 3.2 and 3.3) would spell out the purpose, intended outcomes, whom the process is intending to serve, what implementation looks like, and the results. All of these tools should make it easy to know what 100-percent implementation looks like. The results shown in the monitoring tool will indicate the degree to which the process is implemented with integrity and fidelity. Questionnaire results will help us understand what teachers and students are experiencing with process implementation.

**FIGURE 4.2**

**Sample Process Measurement Planning Table**

| *What process do you want to measure?* | *What do you want the process to look like?* | *How can this process be measured?* |
|---|---|---|
| **Curriculum** | Aligned to standards and grade-level expectations; continuum of learning that makes sense to students; interesting; implemented in every classroom. | • Curriculum mapping<br>• Process flowchart<br>• Classroom/teacher observations<br>• Student achievement results (student data and student work)<br>• Staff, student, parent, and standards questionnaires<br>• Vision assessment tool |
| **Instruction** | Agreed-upon strategies implemented in every classroom, including small- and large-group instruction, flexible groupings, differentiated instruction, scheduling; designed to meet the needs of all students. | • Process flowchart<br>• Measures of instructional coherence<br>• Classroom/teacher observations<br>• Student achievement results (student data and student work)<br>• Staff and student questionnaires<br>• Vision assessment tool |
| **Assessments for Learning** | Formative assessments aligned to the standards, grade-level expectations, and high-stakes summative assessments. | • Assessment inventory<br>• Uses of assessments<br>• Process flowcharts<br>• Classroom/teacher observations<br>• Student achievement results (student data and student work)<br>• Staff and student questionnaires<br>• Vision assessment tool |
| **Staff Collaboration** | Teachers meet in teaching teams to review student progress (student data and student work), improve the implementation of the vision, and adapt processes. | • Staff questionnaire<br>• Vision assessment tool<br>• Leadership structure<br>• Instructional improvement |

| *What process do you want to measure?* | *What do you want the process to look like?* | *How can this process be measured?* |
|---|---|---|
| **Learning Environment** | Students feel like they belong, are challenged, and are cared for.<br>Teachers feel supported and that they are working in a collaborative environment; teachers have high expectations for students and believe all students can learn.<br>Parents feel welcome at the school, and know what they can do to support their child's learning; effective home-to-school communications. | • Student, staff, and parent questionnaires<br>• Demographic data that indicate how students and staff are treated, in regard to behavior, attendance, retention, special education, and gifted. |
| **Leadership** | Leadership structure that helps everyone implement the vision; supportive of all staff, students, and parents; supports the continuous improvement of the organization and all personnel. | • Student and staff questionnaires<br>• Leadership structure that helps everyone implement the vision<br>• Evaluation tools and strategies |
| **Sustainable Learning Organization** | Professional learning that is sustained and supports individuals in implementing the vision. | • Demographic data<br>• Organizational student, staff, parent questionnaires<br>• Processes that make sense for students |
| **Integrity and Fidelity** | Integrity is implementing instruction or a program with accuracy and consistency.<br>Fidelity is implementing instruction or a program the way it is intended to be implemented. | • Implementation monitoring tool<br>• Program Evaluation Tool completed for each program or process to be implemented<br>• Observations<br>• Process flowchart<br>• Student engagement<br>• Student questionnaires |

## Evaluating Processes with Flowcharting

Process flowcharts are a means of measuring what we want a process to look like. A process flowchart is an extremely helpful tool for showing how a program is intended to be implemented and how it is currently being implemented, which is helpful for improving process or program implementation.

By flowcharting the processes being implemented, schools can clarify what is being done now, so all those involved can understand how they are getting current results and determine what needs to change to get different results. A flowchart allows everyone to see the major steps in a process, in sequence, and then evaluate the differences between the theoretical and actual, or actual and desired, results. The intent is to agree on standard and desired practice, and then improve the current processes. A flowchart, or process map, is a visual representation of a process that helps staff:

- assess what is really being implemented within a process;
- understand how they get the results they are getting;
- determine the cause of a problem or challenge;
- improve a process; and
- train and communicate process expectations so all staff can understand and implement the same effective process.

A flowchart can be constructed both informally and formally. An informal method is best for getting started and securing staff engagement; the formal method ensures rigor and accuracy. A well-prepared flowchart:

- builds common understandings of the whole process (it is best if staff work together to create the flowchart to realign any misconceptions along the way);
- communicates process-related ideas, information, and data in an effective visual form;
- identifies actual or ideal paths, revealing problem areas and potential solutions;
- highlights areas for improvement;

- breaks processes into steps using consistent, easily understood symbols;
- is inexpensive and quick to produce, and gives staff the opportunity to experience a shared view when they participate in constructing it;
- shows intricate connections and sequences clearly;
- aids in communication, problem solving, and decision making;
- promotes understanding of a process in a way that written procedures cannot;
- enables the standardization of a process; and
- provides a way to monitor and update processes.

One good process flowchart can replace pages of words. A flowchart by itself is qualitative data. Data that test the theories inherent in the flowchart make the flowcharting process quantitative.

## Flowcharting Your Processes

A large district I was working with wanted me to teach principals how to flowchart processes, as a part of a larger workshop. After some instruction, I asked them to pick any process to flowchart on chart paper so we could do a gallery walk when everyone was finished. It was not long before we could see that almost all the principals were choosing to flowchart Professional Learning Communities. The district administrator working with me wanted me to ask the principals to flowchart anything but Professional Learning Communities. I asked why. Her response was, "Because they have had three years of professional development in Professional Learning Communities—they should know what it looks like!" I asked her to trust me and allow the principals to proceed.

I am guessing you know what happened next: No two flowcharts looked alike. Some principals reported, quite honestly, that they told their teachers that those grade-level teams are now Professional Learning Communities, or that staff meetings are now Professional Learning Communities, if anyone asks. On the other end, some principals had elaborate flowcharts that showed the intricacies of teachers reviewing data together and supporting one another in the classroom to change practices. The big learners of the day were the district administrators. They realized the schools did not need more professional learning in Professional Learning Communities. They needed guidance. They also realized that by being flexible and not clarifying an approach, they would not get what they wanted as intended outcomes for the district. They had to be more direct, while also being flexible. In this case, they created a simple flowchart that gave the components they wanted to see in each school's Professional Learning Communities. The schools could put the components together in whatever manner they wanted. Figure 4.3 shows this simple flow of ideas.

Before flowcharting a process, staff need to answer these questions:

- What is the purpose of this process flowchart?
- What are the intentions of the process?
- What are the desired results?
- How will we know if the process is being implemented?

These are essentially the same questions asked in the Program Evaluation Tool. These two work very well together.

Basic flowcharting symbols include the following:

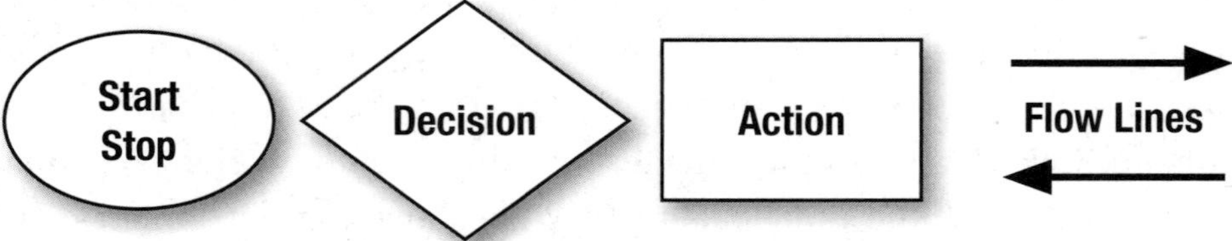


Flowcharts are useful tools for exposing problems and incomplete thinking in processes. Improvement cannot occur until problems are identified and solutions are proposed. This requires gathering data on student achievement results and on the processes used to produce these results. In addition, incorporating data on demographics and perceptions can help us acquire a true picture of which processes are working with which students. For example, questionnaire data, disaggregated by demographics, could show one student group feeling disenfranchised about a particular

FIGURE 4.3

**The Process of Using Data in Professional Learning Communities Flowchart**

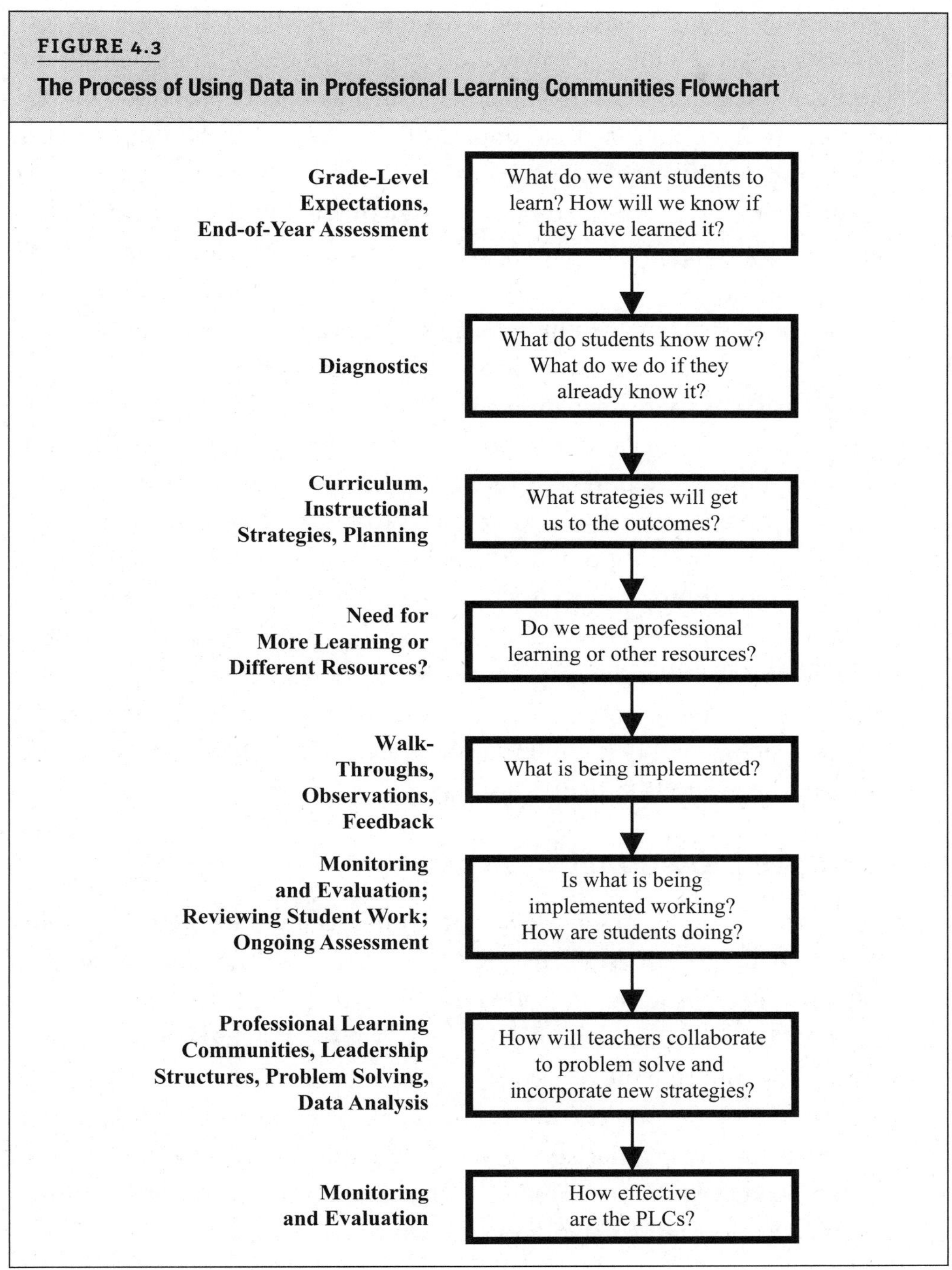

process in a couple of classrooms, or one or more student groups feeling challenged by the work they are asked to do when certain conditions apply. Additionally, demographic data, such as attendance and behavior, could give us insight into the impact of implementing specific strategies. We cannot just assume that we know which processes work best, or that we have a common process; we must analyze the instructional strategies that are actually being implemented and determine which ones are getting the desired results.

Some schools have found it helpful to have individuals, or teams of individuals, create their own flowcharts of how a process works, before working as a larger team. Coming together with very different pictures can be a very important "aha" for a school. It can reveal weak understanding or weak implementation, and direct a fix. For example, five small groups in one school I worked with were asked to draw a flowchart of the same reading program. The group drew five different flowcharts that represented five different approaches to teaching the same reading program. Some of the information was terrific to help teachers differentiate instruction; however, it was clear that the teachers needed to go back to the purpose of the program and the recommended structure. Together, they created a system of teaching reading that embraced the intent of the reading program and took it to new levels to reach all students.

## Sample Flowcharts

Figures 4.4 and 4.5 present flowcharts of the scenario from the beginning of the chapter that described Day Creek Elementary School's struggle to get all students reading on grade level by grade 3 (with Figure 4.4 outlining Day Creek's original approach to the Reading Program Placement and Figure 4.5 outlining the revised process).

Just about any process can be flowcharted, including a school's leadership structure/shared decision-making structure, Response to Intervention system, and even the school's vision. Figure 4.6 shows a flowchart created for the District Mathematics Program described in Figure 3.3.

**FIGURE 4.4**

**Day Creek Elementary Grade 1 Reading Program Placement: Original Approach**

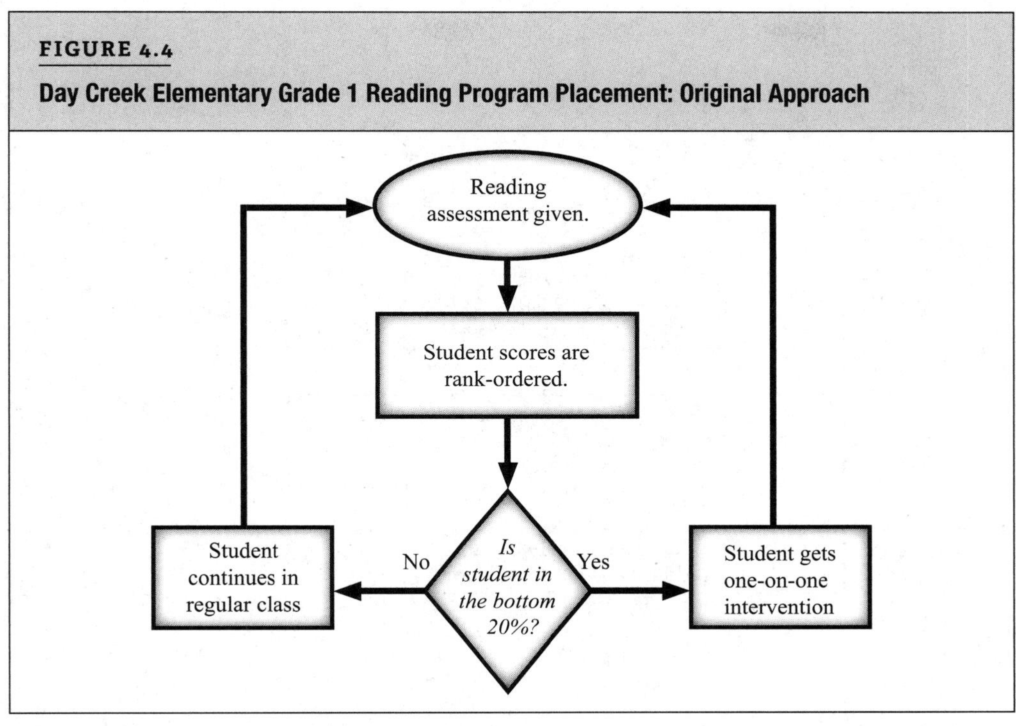


**FIGURE 4.5**

**Day Creek Elementary Grade 1 Reading Program Placement: Revised Approach**

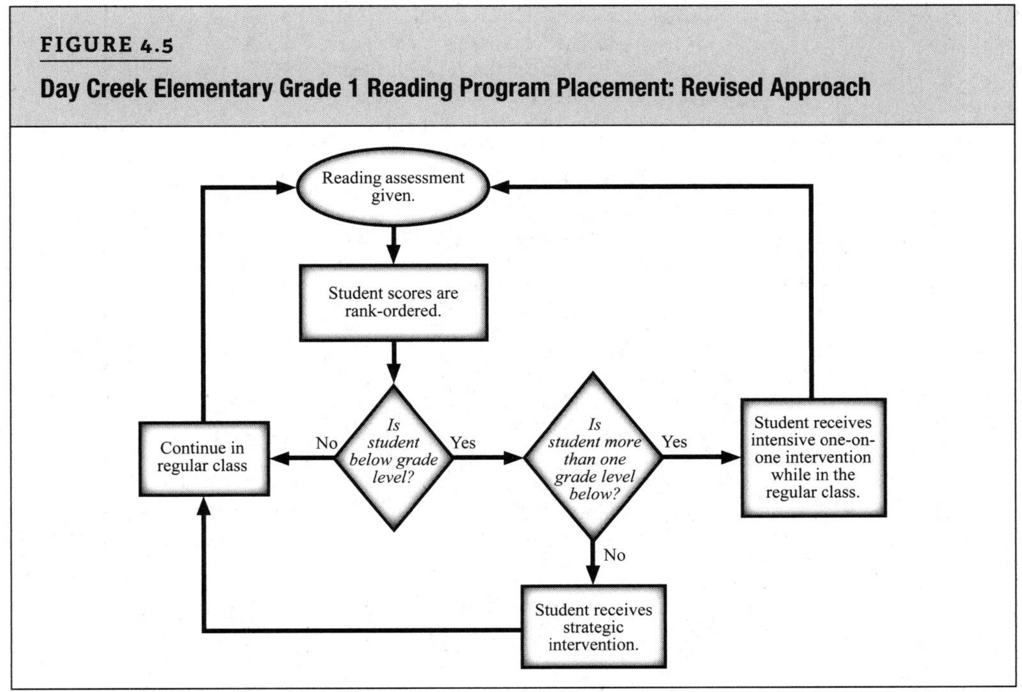

**FIGURE 4.6**

**District Mathematics Program Flowchart**

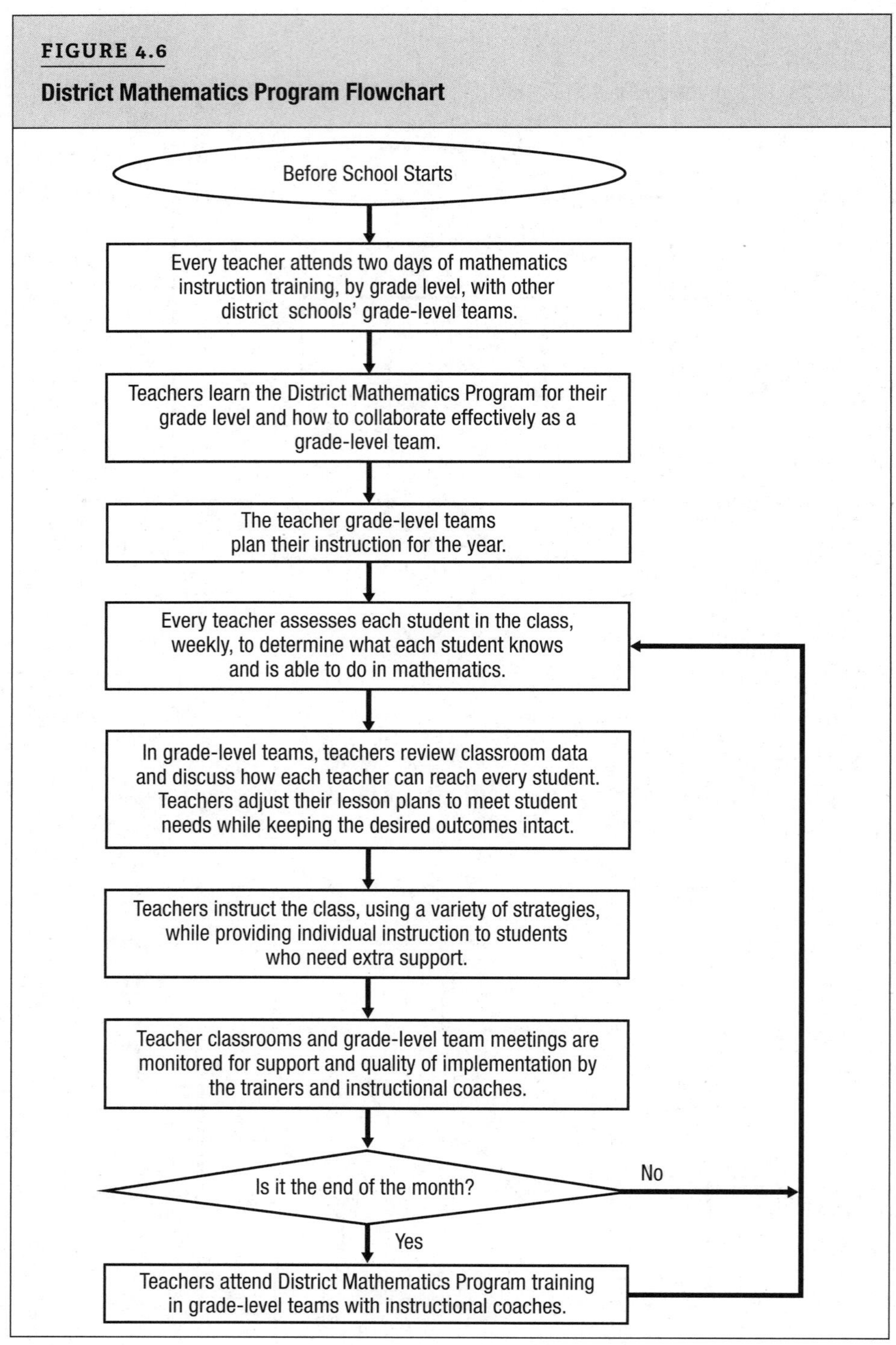

Even the hardest-to-measure processes can be evaluated with the Process Measurement Planning Table, process flowcharting, and by thinking about what evidence you have to measure implementation. Next, we pull it all together to learn how to use the evaluations of these processes, along with programs, to evaluate the entire school.

# 5

# How Can We Evaluate the Entire Learning Organization?

*You've got to think about big things while you're doing small things, so that all the small things go in the right direction.*

—Alvin Toffler

K Street Middle School used the Program Evaluation Tool to evaluate each of its programs and processes, as a part of its continuous school improvement efforts. The tool, along with the Process Measurement Planning Table and process flowcharts, helped the school get everyone on the same page in implementing its programs and processes, and set up each program and process to be monitored and evaluated. What was missing was how the programs and processes worked together to create the whole, and how to determine what to adjust along the way to create a system that achieved the intended outcomes. After creating and committing to a new mission and shared vision, staff asked if they could analyze their new mission and vision the same way they did their programs and processes, and be able to determine if they were on target to achieve the results they wanted. We set out to do just that, and more.

The first thing we did was complete a Program Evaluation Tool for the mission and vision. Figure 5.1 presents the first pages of the K Street Middle School vision in the Program Evaluation Tool. The purpose of the school is reflected in its mission. The vision—the way the mission is implemented—is summarized in the implementation columns, and in an accompanying flowchart (not shown here).

**FIGURE 5.1**

**K Street Middle School Vision**

| ***Needs Assessment*** | ***Purpose*** | | ***Participants*** | ***Implementation*** | | ***Results*** | |
|---|---|---|---|---|---|---|---|
| *What are the data telling about the need for the program or process?* | *What is the purpose of the program or process?* | *What are the intended outcomes?* | *Whom is the program/process intended to serve?* | *How should the program be implemented to ensure attainment of intended outcomes?* | *How is implementation being monitored?* | *How will results be measured?* | *What are the results?* |
| The K Street Middle School Comprehensive Needs Assessment showed the following:<br>• There is a sizeable gap (70%) between our highest-performing students and our lowest-performing students.<br>• Not all students are showing learning growth.<br>Proficiency on the State Assessment is as follows:<br>• ELA: 15% proficient<br>• Math: 17% proficient<br>• Science: 12% proficient | The purpose of K Street Middle School is to take students from elementary school, guide them through the middle school years with high-quality instruction and caring, and provide them with the tools to be successful in high school, in college, and in their careers. | When the purpose of K Street Middle School is implemented as intended:<br>• Instructional coherence and a continuum of learning that makes sense for all students will be evident.<br>• What students learn in one grade level will build on what they learned in previous grade levels, and prepare them for the next grade level. | K Street Middle School is intended to serve all students who come through the doors.<br>***Who is being served? Who is not being served?***<br>The students who are proficient are being served. The students who are not proficient are not being served. | • Before school starts, teachers in their subject-area teams determine what they are going to teach, and the materials they are going to use, across the three grade levels so there is continuity in learning. | The Program Evaluation Framework is used to monitor the implementation of the major programs and processes.<br>The vision implementation tool is used as a teacher self-assessment tool, coupled with an outside review by the curriculum coordinator and instructional coaches. | Student, staff, and parent questionnaires will be administered to understand what the stakeholders are thinking and feeling about the school.<br>Student learning growth and proficiency will be measured through the State Assessment and through common formative assessments. | At the end of the first year of implementing our new vision, we saw more students becoming proficient, with more students showing at least a year's growth in learning.<br>Our RtI system got started with some hiccups, but will be working well by next year.<br>Student attitudes about school have improved over the past two years. Students are feeling challenged. |

*continued*

**FIGURE 5.1** *(continued)*

## K Street Middle School Vision

| ***Needs Assessment*** | ***Purpose*** | | ***Participants*** | ***Implementation*** | | ***Results*** | |
|---|---|---|---|---|---|---|---|
| *What are the data telling about the need for the program or process?* | *What is the purpose of the program or process?* | *What are the intended outcomes?* | *Whom is the program/process intended to serve?* | *How should the program be implemented to ensure attainment of intended outcomes?* | *How is implementation being monitored?* | *How will results be measured?* | *What are the results?* |
| Attendance: While the overall attendance rate is 94%, approximately 12% of the students are chronically absent.<br>Behavior: There are more behavior incidences the higher the grade level.<br>Attitudes:<br>• Not all students feel they are listened to by their teachers.<br>• Students in upper grade levels do not feel challenged by the work they are asked to do.<br>• Students feel other students' behavior gets in the way of their learning. | *The mission of K Street Middle School is to guide all students across their bridge to success by providing them with the support and skills they will need to live in an ever-changing world.* | • Individual student achievement results will improve each year.<br>• All students will be proficient in all areas. No student will need to be retained.<br>• Progress monitoring and common formative assessments conducted within the classroom will be utilized to identify struggling students and why they are struggling. | | • These same teams meet regularly throughout the year to plan and adjust their instruction on the basis of student assessment results.<br>• All students are screened within two weeks of school starting in their core classroom, and again at the middle and end of the year. | ***How should implementation be monitored?***<br>We are monitoring as described above; however, we want to monitor more often next year.<br>***To what degree is the program being implemented with integrity and fidelity?*** | Teacher effectiveness will be measured with walk-throughs by administration and implementation monitoring by peers.<br>All major programs or processes will be monitored for implementation integrity and fidelity on a continuous basis to ensure the highest quality. | |

| **Needs Assessment** | **Purpose** | | **Participants** | **Implementation** | | **Results** | |
|---|---|---|---|---|---|---|---|
| *What are the data telling about the need for the program or process?* | *What is the purpose of the program or process?* | *What are the intended outcomes?* | *Whom is the program/process intended to serve?* | *How should the program be implemented to ensure attainment of intended outcomes?* | *How is implementation being monitored?* | *How will results be measured?* | *What are the results?* |
| • Teachers feel student behavior needs to improve.<br>• Teachers do not work together to create a continuum of learning.<br>• Not all teachers believe that using data will improve student learning.<br>• Teachers say they believe all students can learn. | | • Interventions matched to student needs will result in student learning increases for *every* student.<br>• All students at risk of low achievement are identified early and "failure" is prevented.<br>• Fewer students will be identified for special education. | | • The results of the screeners are reviewed by the RtI team to get a schoolwide view of the performance of all students, to set cut scores, and to allocate support. | In our first year, staff did a fabulous job of implementing everything they were asked to implement. In our second year, we are hoping the implementation will become the way we naturally do business. | | |

*continued*

**FIGURE 5.1** *(continued)*

**K Street Middle School Vision**

| Needs Assessment | Purpose | | Participants | Implementation | | Results | |
|---|---|---|---|---|---|---|---|
| *What are the data telling about the need for the program or process?* | *What is the purpose of the program or process?* | *What are the intended outcomes?* | *Whom is the program/process intended to serve?* | *How should the program be implemented to ensure attainment of intended outcomes?* | *How is implementation being monitored?* | *How will results be measured?* | *What are the results?* |
| | | • Students will not be placed in special education for the wrong reasons—such as teachers wanting students out of the classroom because of behavior or lack of learning response, poor test-taking skills, second-language learning, or lack of adequate interventions.<br>• Referrals made for evaluation of special education are accurate. | | | | | |

| **Needs Assessment** | **Purpose** | | **Participants** | **Implementation** | | **Results** | |
|---|---|---|---|---|---|---|---|
| *What are the data telling about the need for the program or process?* | *What is the purpose of the program or process?* | *What are the intended outcomes?* | *Whom is the program/process intended to serve?* | *How should the program be implemented to ensure attainment of intended outcomes?* | *How is implementation being monitored?* | *How will results be measured?* | *What are the results?* |
| | | • Attendance and student engagement will improve because we are meeting the needs of every student.<br>• Behavior will improve because we are engaging all students in learning. | | | | | |

**Implications for the Continuous Improvement Plan:**

We all need to be responsible for making sure every staff member implements the vision with integrity and fidelity.

Our next challenge was to establish the data needed to understand where the school was, the results it was getting, and the results staff wanted in the short-term, intermediate, and long-term. The K Street Middle School Process Measurement Planning Table is shown as Figure 5.2. This work helped every teacher understand how to think about the process pieces, that the processes would be measured, and how they would be measured. We know these are excellent ways to get everyone on the same page and implementing with integrity and fidelity.

## The Logic Model

The Logic Model helped us lay out the entire system with all data considerations. To effectively evaluate a learning organization, as with a program or process, the evaluation team must study the system as completely as possible. One of the most effective ways to understand the system, its contexts, and desired outcomes, as well as to design a plan for the evaluation, is to use a Logic Model—which draws a picture of a learning organization's Theory of Change to help it better serve its students. Theory of Change describes the linkages of system operations—how elements work together to help the learning organization plan and carry out efforts to achieve intended outcomes, fulfilling the organization's purpose, and, ultimately, realizing its vision. The Logic Model provides the visualization of how these parts fit together to create the whole, given the inputs. In effect, the Logic Model displays how we do school; it describes a strategy for setting up a program with inputs, programs, processes, and outcomes, and spells out the data required to test the model using formative and summative evaluations.

The Logic Model, made popular by the W.K. Kellogg Foundation in 1998 to help program managers understand how to improve their programs to help mankind, has been adapted in this book for the purposes of schoolwide evaluation. A summary of a simple Logic Model, presented in Figure 5.3, outlines that given the inputs, these are the processes and programs that the learning organization believes will produce the desired short-term, intermediate, and long-term outcomes. We will now learn how to create a Logic Model, using as an example how K Street Middle used the Logic Model.

**FIGURE 5.2**

**K Street Middle School Process Measurement Planning Table**

| ***What process do you want to measure?*** | ***What do you want the process to look like?*** | ***How can this process be measured?*** |
|---|---|---|
| **Curriculum** | Aligned to standards and grade level expectations; continuum of learning that makes sense to students; interesting; agreed-upon across the three grade levels. | • Curriculum mapping<br>• Classroom/teacher observations<br>• Student achievement results (student data and student work)<br>• Staff, student, parent, and standards questionnaires<br>• Instructional coherence |
| **Instruction** | Agreed-upon strategies implemented in every classroom, including small- and large-group instruction, flexible groupings, differentiated instruction, scheduling; designed to meet the needs of *every* student. | • Measures of instructional coherence<br>• Classroom/teacher observations<br>• Student achievement results (student data and student work) builds from one year to the next<br>• Staff and student questionnaires<br>• Vision assessment tool |
| **Assessments** | Formative assessments aligned to the standards, grade-level expectations, that measure interventions, and high-stakes summative assessments. | • Assessment inventory<br>• Uses of assessments<br>• Process flowchart for RtI system<br>• Classroom/teacher observations<br>• Student achievement results (student data and student work)<br>• Staff and student questionnaires<br>• Vision assessment tool |
| **RtI** | An instruction and assessment system that seamlessly provides the learning forum that accelerates each student's learning. All teachers and staff working together to ensure every student's success. | • RtI system flowchart<br>• Interventions that are accelerating each student's learning, while keeping each student in his/her regular class<br>• RtI assessment results<br>• Grades, student work |

*continued*

**FIGURE 5.2** *(continued)*

**K Street Middle School Process Measurement Planning Table**

| ***What process do you want to measure?*** | ***What do you want the process to look like?*** | ***How can this process be measured?*** |
|---|---|---|
| **Staff Collaboration** | Teachers meet in teaching teams to review student progress (progress-monitoring data and student work), to improve the implementation of the vision, and to adapt processes, as needed. | • Staff questionnaire<br>• Vision assessment tool<br>• Leadership structure<br>• Instructional improvement |
| **Learning Environment** | Students come to school and feel like they belong, are challenged, and are cared for.<br>Teachers feel supported and that they are working in a collaborative environment; teachers have high expectations for students and believe every student can learn.<br>Parents feel welcome at the school, and know what they can do to support their child's learning; effective home-school communications. | • Student, staff, and parent questionnaires<br>• Behavior data<br>• Teacher and student attendance |
| **Leadership** | Leadership structure that helps everyone implement the vision; supports all staff, students, and parents; supports the continuous improvement of the organization and all personnel. | • Student and staff questionnaires<br>• Leadership structure that helps everyone implement the vision<br>• Evaluation tools and strategies |
| **Sustainable Learning Organization** | Professional learning that is sustained, supports individuals in implementing the vision, and helps each teacher become better. | • Demographic data<br>• Organizational student, staff, parent questionnaires<br>• Processes that make sense for students |

| *What process do you want to measure?* | *What do you want the process to look like?* | *How can this process be measured?* |
|---|---|---|
| **Integrity and Fidelity** | Integrity is implementing instruction or a program with accuracy and consistency.<br>Fidelity is implementing instruction or a program the way it is intended to be implemented. | • Vision implementation monitoring tool<br>• Program Evaluation Tool completed for each program or process to be implemented<br>• Observations<br>• Process flowchart<br>• Student engagement<br>• Student questionnaires |

## Creating Your Logic Model

The steps for creating and testing your own Logic Model are as follows:

1. *Determine what you want to illustrate with your Logic Model.* How do you want to describe the vision for your learning organization or a program? K Street Middle School wanted to show its vision in a Logic Model, how all the parts worked together, and what data the school needed to use to determine overall effectiveness.
2. *Start with the end in mind.* Determine the expected short-term and long-term outcomes for the program or vision you are describing. You can start from scratch or do this by adapting an existing model to fit your school's Theory of Change. If you have a Program Evaluation Tool already completed, the outcomes and how you want to measure them are already thought through. Intermediate outcomes are also an option. K Street Middle School used its continuous school improvement data analysis work and Program Evaluation Tool for its vision to establish inputs, processes, and outcomes. Figure 5.4 shows the short-term, intermediate, and long-term outcomes. (Staff enjoyed listing the outcomes and discussing which should be a short-term, intermediate, and long-term.)

**FIGURE 5.3**

**Summary of Simple Logic Model**

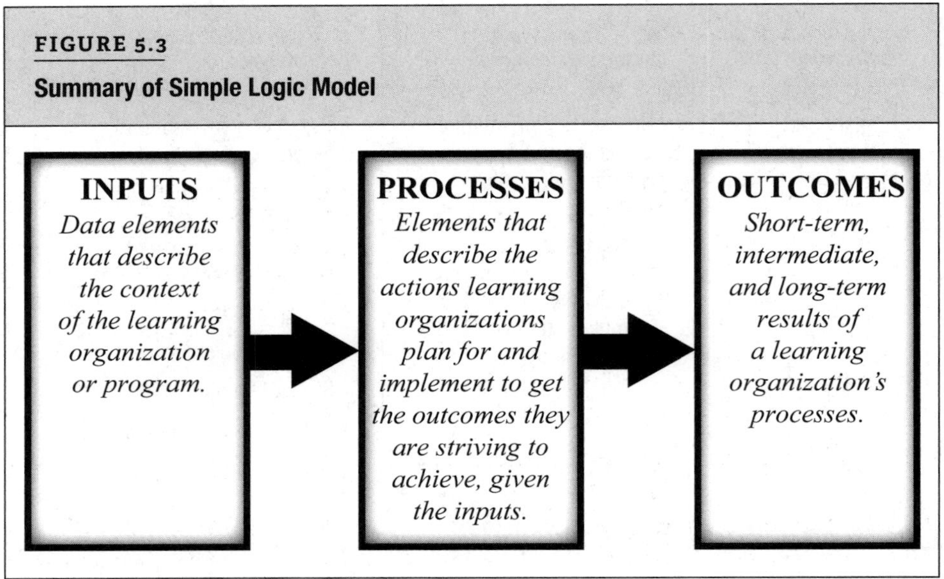


3. *Lay out the inputs—demographics, student learning, perceptions, and school processes.* List the data that must be considered as your school adapts its programs, processes, and procedures to meet the needs of *all* students. Inputs describe where the school is now. K Street Middle School was fortunate to have a comprehensive data profile of its demographics, perceptions, student learning, and school processes data through its continuous school improvement work. The demographic and student learning profiles (organized from most general information to more specific information, over time) are provided in the State Longitudinal Data System. The school added questionnaire results and school processes analysis. For inputs, school staff thought in terms of the data they have for their school, students, teachers, principals, and parents. Figure 5.5 presents the inputs that describe the context of K Street Middle School.
4. *List high-quality programs and processes.* List the programs and processes your staff believe everyone should be implementing with integrity and fidelity to achieve the short-term, intermediate, and long-term outcomes, given the inputs. K Street Middle School Staff made a list of their programs and processes during their continuous

**FIGURE 5.4**

**K Street Middle School Expected Outcomes**

| ***Short-Term*** | | ***Intermediate*** | ***Long-Term*** |
|---|---|---|---|
| *Students*<br>• Ability to plan for the future<br>• Able to compute<br>• Able to read<br>• Able to think critically<br>• Able to think scientifically<br>• Able to use technology effectively<br>• Able to write effectively<br>• Appreciation of the arts and music<br>• Feeling that teachers care about them<br>• Feeling that they belong<br>• Good attendance<br>• Involved in extracurricular activities<br>• Involved in school activities<br>• Knowledgeable of history<br>• Leadership<br>• Listening<br>• Logical Reasoning<br>• Love school<br>• Positive attitudes about learning<br>• Positive behavior<br>• Positive grades<br>• Positive self-esteem<br>• Problem solving<br>• Speaking/listening<br>• Student learning growth every year | *Administration and Staff*<br>• Caring relations with students<br>• Collaboration with each other<br>• Engaged in partnerships with parents, business, and community<br>• Good morale<br>• Implementing quality instruction that meets the needs of students<br>• Program alignment to high school, college, and workforce.<br>• Providing a physically and emotionally safe environment<br>• Satisfied with services they are offering<br>• Understanding the impact of their actions and policies on students<br>• Using data to guide decision making<br><br>*Parents/Families*<br>• Engaged in partnerships with teachers, principals, and students<br>• Involved with child's learning<br>• Knowledgeable of strategies to enhance children's learning<br>• Satisfied with services the school provides | *Students*<br>• Ability to plan for the future<br>• Employment<br>• Enrollment in college—two-year, four-year<br>• High school graduation<br>• Involved in extracurricular activities<br>• Involved in school activities<br>• Leadership<br>• No remedial courses in college<br>• No repeat or make-up courses in high school<br>• Positive grades<br>• Scholarships<br>• Successful in work<br>• Successful SAT/ACT/State Assessment results<br><br>*Administration and Staff*<br>• Effective classrooms<br>• Implementation of vision with integrity and fidelity<br>• Increased instructional skills<br>• Increased teacher confidence in differentiating instruction and grouping students | *Students*<br>• College graduation—two-year, four-year<br>• Community involvement<br>• Successful in college<br>• Successful in life<br>• Successful in work<br><br>*Parents*<br>• Able to support their childrens' lifelong ambitions<br><br>*Community*<br>• Benefiting from having great schools and employees |

**FIGURE 5.5**

**K Street Middle School Inputs**

| *School* | *Students* | *Teachers* | *Principals* | *Parents* |
|---|---|---|---|---|
| • Attendance<br>• Culture/ environment/ safety<br>• Enrollment<br>• Grade-level expectations<br>• Mobility rates<br>• Remediation<br>• Resources<br>• Special Education<br>• Accountability ratings | • Attendance<br>• Attitudes<br>• Behavior<br>• Discipline<br>• Elementary school successes<br>• Extracurricular activities<br>• Gender, ethnicity/race, qualifiers for free/reduced-cost lunch, homeless, first language, second language, special education, gifted<br>• Health<br>• Learning styles<br>• Mobility<br>• Pre-K attendance<br>• Retentions<br>• Tardies | • Gender, ethnicity/race<br>• Qualifications, training, experience<br>• Number of years teaching<br>• Teaching styles<br>• Attitudes<br>• Core values and beliefs<br>• Attendance | • Gender, Ethnicity/race<br>• Qualifications, training, experience<br>• Number of years as administrator/ principal<br>• Leadership styles<br>• Attitudes<br>• Core values and beliefs<br>• Attendance | • Socioeconomic status<br>• Participation in school events<br>• Educational background<br>• Attitudes<br>• Number of children in the home<br>• Number of parents/people in the home<br>• Language spoken at home |

school improvement data analysis work and addressed which programs and processes they wanted to keep during their shared vision work. They listed programs and processes separately in their Logic Model, although there was a lot of discussion about how each program had an important process, and that some processes could also be programs. Understanding the overlap, they decided on the processes and programs presented in Figure 5.6. (This was their first attempt at creating a Logic Model. They will consider merging the two columns next year.)

**FIGURE 5.6**

**K Street Middle School Programs and Processes**

| *Processes* | *Programs* |
| --- | --- |
| • Acceleration processes<br>• Alignment of curriculum, instruction, and assessment<br>• Assessment (common, formative, progress monitoring, grades, standards)<br>• Behavior processes<br>• Business partnerships<br>• Differentiated instruction<br>• Field trips<br>• Grading policies<br>• Guidance process<br>• Homework<br>• Lesson planning<br>• Literacy and subject-area coaching support<br>• Parents/family involvement<br>• Professional learning communities<br>• Professional learning for teachers, support staff, and principals<br>• Program evaluation<br>• Promotion and retention policies<br>• Response to Interventions (RtI)<br>• Standards curriculum<br>• Teacher collaboration<br>• Teaching strategies | • After-school programs<br>• At-risk programs<br>• Athletic programs<br>• Career planning<br>• Career/technology education<br>• Debate and speech competitions<br>• English as a second language<br>• Family resource center/programming<br>• Gifted programs<br>• Guidance program<br>• Homeless program<br>• Interventions<br>• Language programs<br>• Leadership<br>• Music program<br>• Peer mediation<br>• Special education programs<br>• Student teamwork<br>• Study skills program<br>• Transient student program<br>• Transition planning program |

5. *Put it all together and start looking at the linkages.* Lay out the inputs, processes (or processes and programs), and outcomes. Study the linkages with this in mind: given our inputs and the desired outcomes, these are the programs and processes we believe will help us achieve all these outcomes. Figure 5.7 presents K Street middle school's completed Logic Model and Figure 5.8 presents a Logic Model of a typical school.

**FIGURE 5.7**

**K Street Middle School Logic Model**

## Context/Inputs

*School*
- Attendance
- Culture/environment/safety
- Enrollment
- Grade-level expectations
- Mobility rates
- Remediation
- Resources
- Special Education
- Accountability ratings

*Students*
- Attendance
- Attitudes
- Behavior/discipline
- Credits
- Elementary school successes
- Extracurricular activities
- Gender, ethnicity/race, qualifiers for free/reduced-cost lunch, homeless, first language, second language, special education, gifted
- Health
- Honors
- Learning styles
- Mobility
- Pre-K attendance
- Retentions
- Tardies

*Teachers*
- Gender, ethnicity/race
- Qualifications, training, experience
- Number of years teaching
- Teaching style
- Attitudes
- Core values and beliefs
- Attendance

## School Processes

*Processes*
- Acceleration processes
- Alignment of curriculum, instruction, assessment
- Assessment (common, formative, progress monitoring, grades, standards)
- Behavior processes
- Business partnerships
- Differentiated instruction
- Field trips
- Grading policies
- Guidance process
- Homework
- Lesson planning
- Literacy and subject-area coaching support
- Parents/family involvement
- Professional learning communities
- Professional learning for teachers, support staff, and principals
- Program evaluation
- Promotion/retention policies
- Response to Interventions (RtI)
- Standards curriculum
- Teacher collaboration
- Teaching strategies

*Programs*
- After-school programs
- At-risk programs
- Athletic programs
- Career planning
- Career/technology education
- Debate and speech competitions
- English as a second language
- Family resource center/programming
- Gifted programs
- Guidance program
- Homeless program
- Interventions
- Language programs
- Leadership

## Outcomes

### *Short-Term*

*Students*
- Ability to plan for the future
- Able to compute
- Able to read
- Able to think critically
- Able to think scientifically
- Able to use technology effectively
- Able to write effectively
- Appreciation of the arts and music
- Feeling that teachers care about them
- Feeling that they belong
- Excellent senior projects
- Good attendance
- Involved in extracurricular activities
- Involved in school activities
- Knowledgeable of history
- Leadership
- Listening
- Logical reasoning
- Love school
- Positive attitudes about learning
- Positive behavior
- Good grades
- Positive self-esteem
- Problem solving
- Speaking
- Student learning growth every year

*Administration and Staff*
- Caring relations with students
- Collaboration with one another
- Engaged in partnerships with parents, business, and community
- Good morale
- Implementing quality instruction that meets the needs of students
- Program alignment to high school, college and workforce
- Providing a physically and emotionally safe environment
- Satisfied with services they are offering

### *Intermediate*

*Students*
- Ability to plan for the future
- Employment
- Enrollment in college—2-year, 4-year
- High school graduation
- Involved in extracurricular activities
- Involved in school activities
- Leadership
- No remedial courses in college
- No repeat or make-up courses in high school
- Good grades
- Scholarships
- Successful in work
- Successful SAT/ACT/State assessment results

*Administration and Staff*
- Effective classrooms
- Implementation of vision with integrity and fidelity
- Increased instructional skills
- Increased teacher confidence in differentiating instruction and grouping students

### *Long-Term*

*Students*
- College graduation—2-year, 4-year
- Community involvement
- Successful in college
- Successful in life
- Successful in work

*Parents*
- Able to support their childrens' lifelong ambitions

*Community*
- Benefiting from having great schools and employees

| Context/Inputs | School Processes | Outcomes: Short-Term | Outcomes: Intermediate |
|---|---|---|---|
| ***Principals***<br>• Gender, ethnicity/race<br>• Qualifications, training, experience<br>• Number of years as administrator/principal<br>• Leadership style<br>• Attitudes<br>• Core values and beliefs<br>• Attendance | ***Programs*** continued<br>• Music program<br>• Peer Mediation<br>• Special Education Programs<br>• Student teamwork<br>• Study skills program<br>• Transient student program<br>• Transition planning program | ***Administration and Staff*** continued<br>• Understanding the impact of their actions/policies on students<br>• Using data to guide decision making | |
| ***Parents***<br>• Socioeconomic status<br>• Participation in school events<br>• Educational background<br>• Attitudes<br>• Number of children in the home<br>• Number of parents/people in the home<br>• Language spoken at home | | ***Parents/Families***<br>• Engaged in partnership with teachers, principals, and students<br>• Involved with child's learning<br>• Knowledgeable of strategies to enhance childrens' learning<br>• Satisfied with services the school provides | |

**FIGURE 5.8**

## Typical School Logic Model

### Context/Inputs

*School/Learning Organization*
- Enrollment
- Attendance
- Mobility rates
- Graduation rates
- Dropout rates
- Special Education
- Accountability ratings
- Grade-level expectations
- Resources
- Culture/environment/safety

*Students*
- Gender, ethnicity/race, qualifiers for free/reduced-cost lunch, homeless, first language, second language, special education, gifted
- Attitudes
- Learning styles
- Retentions
- Pre-K attendance
- Health
- Mobility
- Behavior/discipline
- Attendance
- Tardies
- Extracurricular activities

*Teachers*
- Gender, ethnicity/race
- Qualifications, training, experience
- Number of years teaching
- Teaching style
- Attitudes
- Core values and beliefs
- Attendance
- Retention rate

### Processes

*Processes and Programs*
- Curriculum
- Instruction
- Assessment (common, formative, grades, standards)
- Alignment of curriculum, instruction, assessment
- Programs (after-school, tutoring, reading, etc.)
- Interventions
- Response to Interventions (RtI)
- Services Learning
- Special Education program
- Gifted programs
- Extracurricular activities
- Dropout prevention
- Environmental learning
- Guidance
- Behavior program/processes
- Language programs
- Healthy breakfasts
- Healthy lunches
- Healthy activities
- Work training
- Study skills
- Student teamwork
- Advanced courses
- Career/technology education
- Career planning
- Teacher observations
- Lesson planning
- Teaching strategies
- Teacher collaboration
- Leadership
- Vision
- Professional learning for teachers, support staff, and principals
- Professional learning communities
- Literacy and subject-matter coaches
- Promotion/retention policies
- Business partnerships
- Use of resources
- Transportation
- Family resource center

### Outcomes

#### *Short-Term*

*Students*
- Graduation
- Employment
- No remedial courses in college
- Positive attitudes about learning
- Enrollment in college—2-year, 4-year
- Successful SAT/ACT/State assessment results
- Ability to plan for the future
- Leadership
- Feeling that they belong
- Love school
- Feeling that teachers care about them
- Positive behavior
- Good attendance
- Positive self-esteem
- Dropout recovery
- Good grades
- Community involvement
- Involved in extracurricular activities
- Involved in school activities
- Scholarships
- Able to write effectively
- Able to think scientifically
- Able to think critically
- Able to read
- Able to compute
- Speaking
- Listening
- Problem solving
- Logical reasoning
- Knowledgeable of history
- Able to use technology effectively
- Appreciation of the arts and music
- Ability to integrate subjects

*Principals and Teachers*
- Satisfied with the services they are offering
- Good morale
- Program alignment to college and workforce
- Caring relations with students
- Able to understand the impact of their actions/policies on students

#### *Intermediate*

*Students*
- College graduation—2-year, 4-year
- Community involvement
- Successful in college
- Successful in life
- Successful in work

*Parents*
- Able to support their childrens' lifelong ambitions

*Community*
- Benefiting from having great schools and employees

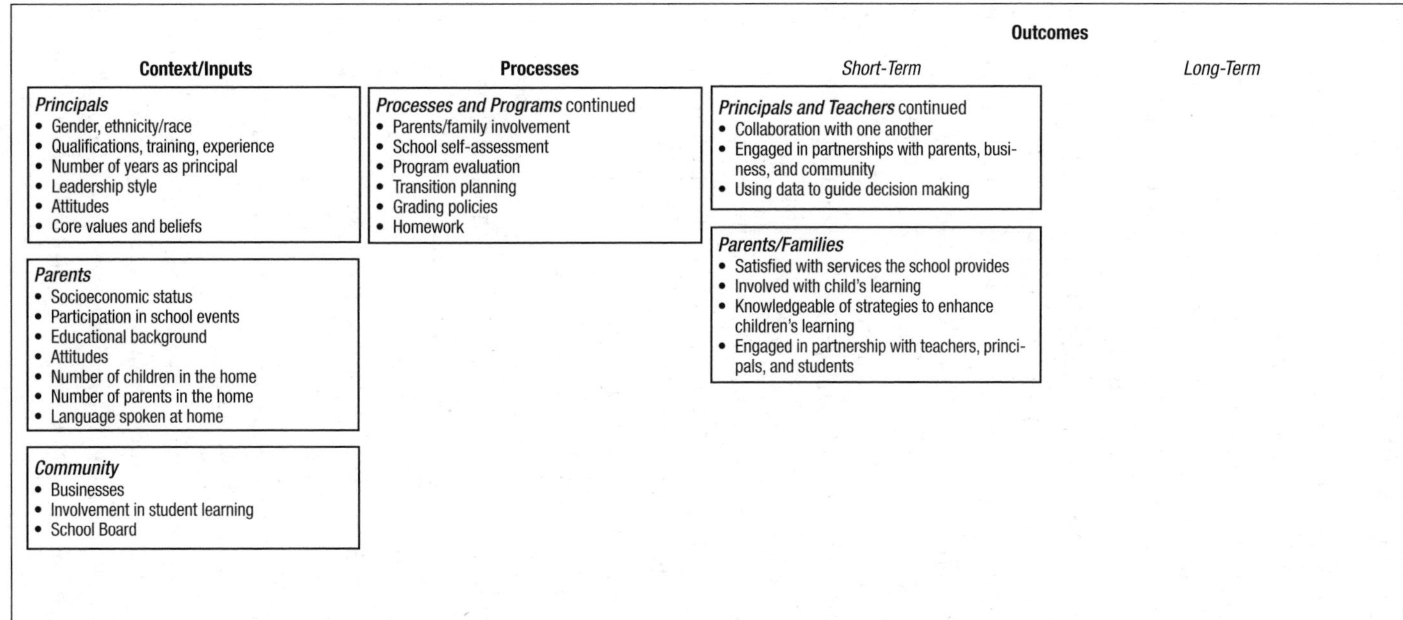
Outcomes
Context/Inputs
Processes
Short-Term
Long-Term
Principals
• Gender, ethnicity/race
• Qualifications, training, experience
• Number of years as principal
• Leadership style
• Attitudes
• Core values and beliefs
Parents
• Socioeconomic status
• Participation in school events
• Educational background
• Attitudes
• Number of children in the home
• Number of parents in the home
• Language spoken at home
Community
• Businesses
• Involvement in student learning
• School Board
Processes and Programs continued
• Parents/family involvement
• School self-assessment
• Program evaluation
• Transition planning
• Grading policies
• Homework
Principals and Teachers continued
• Collaboration with one another
• Engaged in partnerships with parents, business, and community
• Using data to guide decision making
Parents/Families
• Satisfied with services the school provides
• Involved with child's learning
• Knowledgeable of strategies to enhance children's learning
• Engaged in partnership with teachers, principals, and students

## Testing Your Logic Model

The Logic Model can seem overwhelming even before we think about looking at all the data implied within it. However, if a school is working on continuous school improvement, it has much of the data work done already. Inputs describe *Where are we now?* The school would have gathered its multiple measures of data, described in Chapter 2, such as demographics, perceptions, school processes, and student learning data, which represent the inputs in the Logic Model. The update of the data in the Logic Model becomes the outcomes. For example, student attendance is an input (baseline) that we would also want to see improve, as an outcome, or result of our programs and processes. In the Logic Model, programs and processes can appear in their own respective columns, like K Street Middle School did, or as one column combined, as shown in the Typical School's Logic Model. Understood in the Logic Models shown here is that the Program Evaluation Tool has been used, or will be used, along with flowcharts, to evaluate each program and process and to ensure implementation with integrity and fidelity, and to ensure that the programs and processes that make a difference are the ones implemented.

Keep the following in mind when testing your logic model:

1. *Analyze the data to test the logical linkages among "inputs," "processes," and "outcomes," as well as the relationships of the inputs and outcomes to one another*. In addition to its data profile, the State Longitudinal Data System allowed K Street Middle School to analyze data through ad hoc queries. Staff got involved in brainstorming the questions they wanted to answer, with data, about their system and data linkages. Some of the questions included:

    - How are student attitudes, behaviors, backgrounds, and attendance related to student achievement results?
    - Are there students with similar backgrounds and abilities who have different attitudes, behaviors, attendance, and student achievement results because of the way they are treated or because of a program in which they are enrolled?

- Which programs and instructional strategies are making the biggest difference in each student's learning?
- What are the differences in student learning results by teacher effectiveness?
- Do the students who did well in elementary school do well at our middle school?
- Do students who do well at our school do well at high school and then post-secondary education?
- What percentage of our students graduate from high school and then enroll in post-secondary education? How many of our students drop out of high school? What are the characteristics of the students who drop out?

2. *Adjust your Logic Model on the basis of the information acquired through data. Make sure the Logic Model stays alive by ensuring that new programs are included and ineffective programs are deleted, and that the programs and processes are meeting the needs of the students.* After creating its vision and using the Program Evaluation Tool for its programs and processes, the initial Logic Model the K Street Middle School put together required changes because inputs changed. The school's demographics changed when a new homeless shelter and low-cost housing construction was completed in its attendance area. The number of students living in poverty increased by 35 percent to an 85-percent total for the school, and the number of homeless increased from roughly 20 homeless students, for which the school did little extra, to 100 known homeless students, for whom the school committed to do more. The ethnicities and language needs also changed. After much discussion, staff decided not to change their outcomes. They would strive for the same outcomes for their students, regardless of their backgrounds. To do this, the way that they did things—their programs and processes—had to change. They adjusted their programs and processes to reach the outcomes, given the inputs. With these changes, they realized they had just become a true learning organization. This excitement empowered them into the new school year.

## Summary

The application of the Logic Model as a planning tool allows precise communication about the purposes of an organization, the components of an organization, the intended outcomes of an organization, and how the organization can be evaluated. Once an organization has been described in terms of the Logic Model, relationships of the items to one another and measures of results can be identified. As the inputs change, the actions must change to achieve the desired outcomes. That is becoming a true learning organization.

The Logic Model supports schools in becoming true learning organizations because it leads to common understandings and provides the big picture of the learning organization's "clients," vision, and intended outcomes. It also helps them be nimble with the planning, implementation, and evaluation of the school vision that will meet the needs of the students, and keep a staff aligned with its vision.

# 6

# Getting Started: Your First Steps Toward Becoming a True Learning Organization

*Learning organizations are . . . organizations where people continually expand their capacity to create the results they truly desire, where new and expansive patterns of thinking are nurtured, where collective aspiration is set free, and where people are continually learning how to learn together.*

—Peter Senge, *The Fifth Discipline*

It is past time to refocus our schools on implementing only those strategies that make a difference for the students. To improve programs and processes, we need to improve them within the context of the system and with the support of extraordinary leadership, to ensure that only the programs that are making a difference for students are implemented, and implemented as intended.

In this book, we have journeyed from looking only at student achievement results, to continuous school improvement, to program evaluation, to monitoring implementation, to systemwide evaluation.

We learned that using only student achievement results to improve our schools and results is akin to trying to fix the kids, assuming everything educators do is just fine. It is definitely not seeing the forest for the trees. Our first big leap on the road to becoming a learning organization was to see how the trees create the forest through continuous school improvement.

Our approach to continuous school improvement is the most logical way to get all staff looking at schoolwide data, including the processes that create the results, and then together determining where the school wants to go, and creating the structures and paths to get there. Getting all staff reviewing schoolwide data is a big step in enlisting them in helping to continuously improve the system. In most schools, staff have never reviewed schoolwide data. They see their grade level and the school's student achievement results, but seldom, if ever, see the entire system. In our experience, staff love seeing the schoolwide data, and seeing schoolwide data helps staff become better equipped at knowing what they need to do to support system improvement. Learning organizations require all staff to have full knowledge of the system, and the nimbleness to change when the data indicates it is time. Everyone has to be on board and knowledgeable, sensitive to when change is needed, and able to make those changes.

To do continuous school improvement well, staff need to evaluate and continuously improve everything they do, all the time. They must ensure that all staff understand how to implement all aspects of the school with integrity and fidelity. To do this, staff must understand the purpose and intended outcomes for each program and process. They must know program and process evaluation. Our Program Evaluation Tool, Process Measurement Planning Table, and process flowcharts, all guided by the continuous school improvement framework and synthesized in the Logic Model, provide a user-friendly approach to designing and evaluating programs and processes.

In this book we talk a lot about monitoring program and process implementation. If a school is not monitoring and measuring program implementation, the program probably does not exist. How is monitoring different and the same as evaluation? Figure 6.1 presents the key activities of monitoring and evaluation—which are very similar. They are both designed to ensure that a program or process is being implemented the way it is intended. The biggest difference is that evaluation is more rigorous and strongly tied to collecting evidence to help evaluators more accurately assess impact. For example, in evaluations, often an evaluator is identified, and the types of data to collect are specified. Evaluation does not have to happen until the end of the program; however, good evaluations are ongoing and provide a lot of the same information as monitoring. Monitoring and evaluation can be done in-house; however, staff are more geared to monitoring implementation rather than "evaluating implementation." Evaluation tends to use more data sources; in fact, all related data sources. Monitoring could use the same data sources, but usually does not. So, to get a better sense of the impact that educators have on student success, don't just monitor—evaluate!

**FIGURE 6.1**

**Key Activities of Monitoring and Program and Process Evaluation**

| ***Action*** | ***Monitoring*** | ***Evaluation*** |
|---|---|---|
| *Purpose* | To ensure and support implementation with integrity and fidelity. | To ensure that the program or process is making the desired difference. To identify what is working and what is not working. |
| *When* | Ongoing, during implementation. | Ongoing during implementation (to nudge it in place) and at the end of a time period (to understand impact). |
| *Where* | In the classroom, or wherever implementation takes place. | Wherever implementation is and wherever the data are: classroom, database, questionnaire results. |
| *Who does the monitoring/ evaluation* | School administration, instructional coaches, teachers, or program trainer. | External evaluator, or a team from inside the school, with assistance from instructional coaches and trainers. |
| *Data collected* | Observation data, external and self-assessment of implementation. | All data related to the program or process: classroom observations; self-assessments; student, staff, and parent questionnaires; interviews; student learning data; and data collected about the participation and implementation. |
| *How used* | To improve the implementation of the program or process while the program is operating and to provide implementation information for evaluation. | To improve the implementation of the program or process while it is operating, to improve the program or process for the next cycle, and to determine impact. |

## How to Become a True Learning Organization

To evaluate the system, processes, and programs, schools need to take the following steps (outlined in Figure 6.2):

1. *Become familiar with the continuous school improvement framework.* Refer to it frequently throughout the evaluation process and let it help guide your thinking.
2. *Review your multiple measures of data with all staff so everyone knows exactly whom you have as students, where you are now as a school, and how you are getting your current results.* Use the strengths, challenges, and implications for the school improvement plan protocol to analyze the data. We highly recommend establishing a data profile of your multiple measures of data to use on an ongoing basis.
3. *Look across the implications* to understand the commonalities.
4. *Go deeper into the challenges to understand contributing causes.* The problem-solving cycle is a wonderful way to get your staff thinking through contributing causes and the data needed to be analyzed to understand more about a problem.
5. *Use all this powerful data analysis work to create a vision for what you ultimately want to do for students.* Make sure the vision is clear, shared, and understood in the same way by everyone.
6. *Ensure everyone understands her or his role in the plan to implement the vision,* by using the Program Evaluation Tool to clarify and set up the evaluation of the shared vision, programs, and processes.
7. *Use the Process Measurement Tool and process flowcharts* to clarify and evaluate programs and processes.
8. *Ensure that teachers are reviewing student data and student work together, and working with each other to improve or eliminate strategies that are not working and to implement new strategies to better meet the needs of students.* Leadership and collaborative structures will help your staff implement the vision.
9. *Monitor the implementation of the vision, programs, and processes* that are a part of the vision.

**FIGURE 6.2**

## Steps to Becoming a True Learning Organization

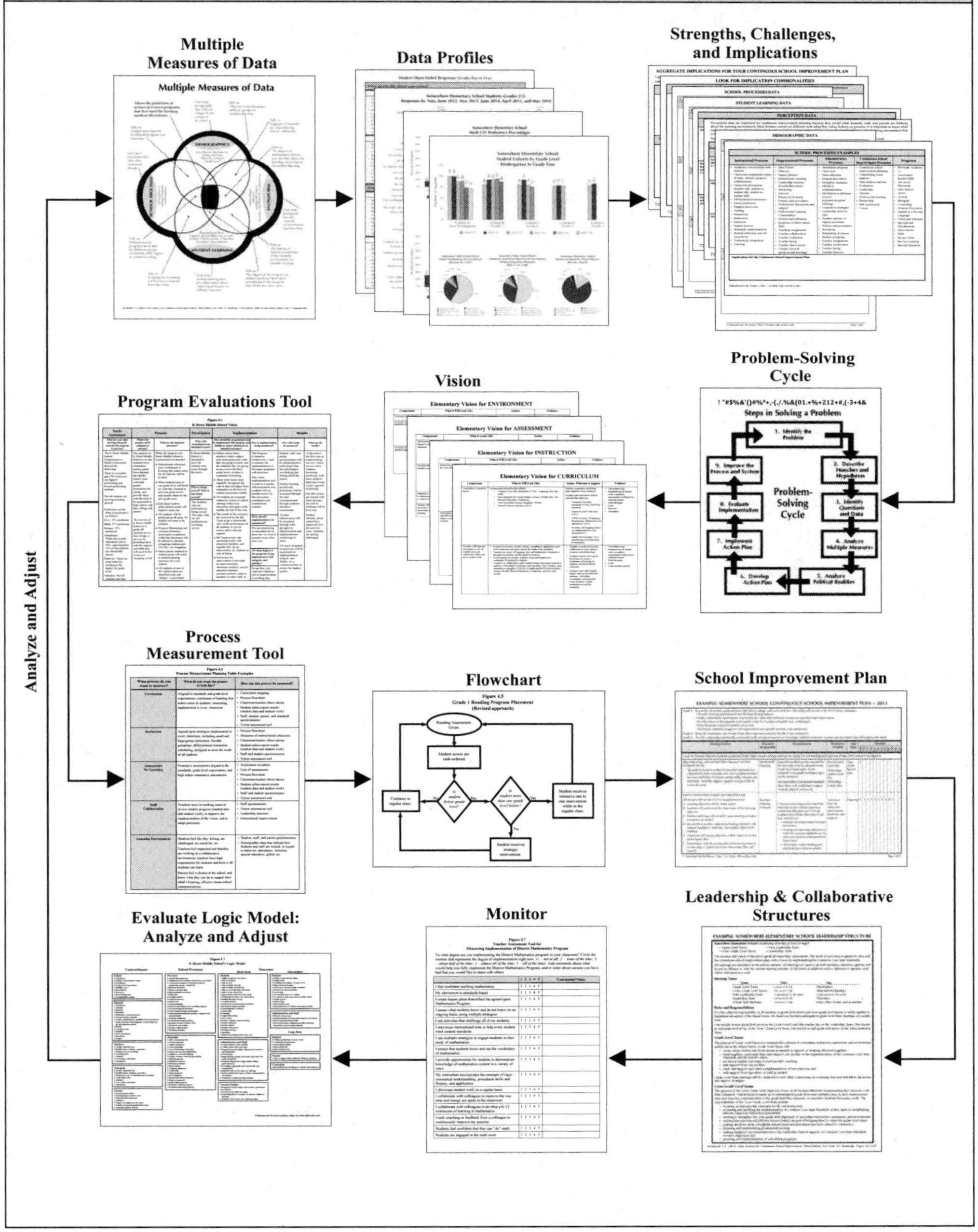

Multiple Measures of Data

Data Profiles

Strengths, Challenges, and Implications

Problem-Solving Cycle

Vision

Program Evaluations Tool

Process Measurement Tool

Flowchart

School Improvement Plan

Leadership & Collaborative Structures

Monitor

Evaluate Logic Model: Analyze and Adjust

Analyze and Adjust

10. *Document the vision through a Program Evaluation Tool, Process Measurement Planning Table, process flowcharts, and Logic Model* to show your Theory of Change.
11. *Use the data and implementation information to analyze and adjust processes, programs, and the shared vision* on an ongoing basis.
    — *Start the cycle again.* Adjust the vision, programs, and processes on the basis of what the data tell you.

If your school is engaged in continuous school improvement, most of this will be in progress already. Continuous school improvement provides opportunities to make school a more engaging place for students to learn and a more satisfying place for teachers to teach.

Evaluating the entire system, as well as all of its parts, provides the opportunities for everyone to continually self-reflect and understand when adjustments need to be made to reach desired results. Reviewing all data, programs, processes, and the shared vision is enlightening to teachers and staff. They can see the big picture. If your school is not engaged in continuous school improvement, there is no time like the present to begin the exciting journey that will make a difference for everyone!

# 7

# Pulling It All Together: A Case Study

For the past three years, Paradise View High School (PVHS) has been working hard to become a true learning organization. The school wants to ensure that *every* student shows learning growth every year because every classroom consists of the highest-quality teaching that meets the needs of every student. Staff want to ensure that their high school is able to know when it is off track and can determine how to get back on track to achieve its desired outcomes. And most of all, they want to achieve all the school's desired outcomes.

When the staff started this journey, PVHS was a focus school in its state. More specifically, it was a school in the bottom 20 percent of the state because it did not have enough students proficient in English Language Arts and Mathematics (as measured on the State's Standards Assessment), had not shown student learning growth for the past two years, and its attendance and graduation rates were too low. The school's leadership team was required to attend training on continuous school improvement. It was during the work they did as a part of the training that led them to become confident that they were capable of becoming a true learning organization—but not without a shared vision, hard work, and some sacrifice.

In the first three days of continuous school improvement training, one of the first things the schools were required to do was to take a hard look at the CSI framework and ask some orienting questions, including:

- Where are we now?
- How did we get to where we are?
- Where do we want to be?
- How are we going to get to where we want to be?
- Is what we are doing making a difference?

The next step was to organize their data to analyze all the school's data—demographics, perceptions, student learning, and school processes. Luckily, the state's data warehouse was set up to provide data profiles, automatically, with the demographic and student learning data for each school in the state. Education for the Future, a not-for-profit initiative I direct out of California State University, Chico that works with educators on systemic change and comprehensive data analyses that lead to increased student learning, provided the student, staff, and parent questionnaires. The challenge was to get every staff member reviewing the schoolwide data to create a plan for improvement.

While skeptical that the staff would agree to analyze data for the whole school, versus just their own classrooms, the leadership team was even more skeptical that the results of the analysis would lead to a plan that staff would want to implement or one that would lead to continuous school improvement. Luckily, the process for reviewing the data made it very easy to get all staff involved and to come to consensus painlessly. The school had had many unsuccessful attempts working with its data in the past. Most attempts turned into what felt like small riots.

## Analyzing the Data

This time, the leadership team facilitated the work, and had the 64 staff members sit in small groups of six or seven, mixed by grade levels and subject areas. The leadership team began the analysis work with demographic data—the data that spelled out the context of the school, and helped everyone see who the students were and some of the life needs they brought to school each day. The data were organized in graphs and charts, over time, from general to more specific—from one source only, so as not to provide confusing or contradictory data. The data read like the story of the school.

In fact, many teachers commented, "This is our story." Each staff member received her or his own copy of the data to read, write on, and keep. All staff members were asked to look at the trends and jot down, as they reviewed the data, what they saw as *strengths, challenges, and implications for the continuous school improvement plan*. Some data can be considered strengths when one can see something positive in the data. *Strength* data can become leverage for improving challenges. Some data imply *challenges* when they show something that might need attention, might be an undesirable result, or something out of a school's control. *Implications for the continuous school improvement plan* are placeholders until all the data are analyzed. *Implications,* or thoughts to address in the continuous school improvement plan, most often result from challenges. It took only 15 minutes for individuals to complete this first phase.

The second phase of looking at the demographic data was to have each small group log on chart paper what they saw collectively in the demographic data with respect to *strengths, challenges,* and *implications for the continuous school improvement plan*. The groups had a great time discussing the data. They realized they were used to talking about strengths and weaknesses, but looking at challenges instead of weaknesses was a huge change. They could see that by identifying challenges, they would be addressing things that *needed* to be addressed in their plan. Many things that were challenges to teachers never showed up in the plan when they just listed weaknesses. For example, the idea that there are many disadvantaged students and more students with learning disabilities never showed up as weaknesses; however, these are challenges to teachers because they must adjust their instruction to meet the needs of the students.

After about 15 minutes of discussions, the small groups posted their collective responses. Through a process of elimination, the small group comments were merged. Individually, staff saw a lot in the data; collectively, they saw so much more. The process brought them to consensus on what needed to be addressed in the continuous school improvement plan, with respect to demographic data—all of this in about 45 minutes. Many wondered out loud why they had not looked at schoolwide data before. Demographic data were making a difference in their minds already. The principal assured them they would be analyzing the schoolwide data every year going forward. Figure 7.1 shows the strengths, challenges, and implications for the continuous school improvement plan that came from the staff analysis of demographic data.

This data analysis process required every staff member to read and understand the data. In the small groups, staff members were validated in what they saw, as others noticed the same things; and they were exposed to new thinking, as some saw things a little differently. As a whole group, staff were amazed at how easy it was to come to consensus, and how "full" the implications were. Even though the data showed that the school was not meeting the needs of its students, staff were excited to continue the process with perceptions data, and then student learning data.

Perceptions data for PVHS include student, staff, and parent questionnaires that provide information about those entities' thoughts about the school learning environment. These are powerful data—as people's actions reflect their values, beliefs, and perceptions. If we want different results, we need to listen to our stakeholders' perceptions. Figure 7.2 shows a summary of the PVHS perceptions data analysis.

As staff reviewed the student learning data, they also created an inventory of all assessments in the school and refined the inventory as much as they could. They then vowed to return to the inventory after the implications of all the data analysis work was complete. Figure 7.3 shows the group summary consensus on two of their student learning assessments—the state standards assessment and the ACT.

When staff got to the school processes data, the approach to data analysis changed a little. During the school process analysis, each staff member received an example list of processes to begin with. As with the other data, individuals had time to review the list, as well as add and cross out instructional, organizational, administrative, continuous school improvement processes, and programs. They quickly merged their thinking. When the staff agreed on the processes and programs operating in the school, they began to analyze them, independently, in small groups, and then the large group, by color coding, as follows:

**FIGURE 7.1**

**Paradise View High School Demographic Data Analysis**

| *Strengths* | *Challenges* |
|---|---|
| • We have a trend of decreasing enrollment over the previous seven years. This year our enrollment is 1,040 students.<br>• Teacher-student ratio is now the lowest ever, with 16.5 students per teacher.<br>• Special education enrollment has decreased over the years, although still high.<br>• The ethnic composition of the student body continues to remain largely unchanged: three major ethnicities, equally distributed.<br>• There has been an increase in daily attendance over the last two years: 89.3% to 90.7%.<br>• The percentage of students who are not suspended has increased (91.7% to 92.5%).<br>• We are not considered a persistently dangerous school.<br>• Property damage and illicit substance offenses have declined from 25 to 17 incidences last year.<br>• Our school has a high four-year college attendance compared to the state.<br>• Our 64 teachers have an average of 10–12 years of teaching. Over half of our teachers have advanced degrees. Six are National Board Certified.<br>• More than 60% of our staff have been at our school for more than five years; teacher turnover has decreased.<br>• 95% of teachers are licensed, with over 90% of our classes being taught by highly qualified teachers. | • As our enrollment decreased over time, we had to cut positions and offerings.<br>• Approximately 130 students left our school each year over the past five years. Most enrolled in private or charter schools.<br>• We still have a large special education population, despite decreasing numbers: 14% (too high).<br>• The data show there might be an over-identification of one ethnicity in special education. Over half of the special education students come from one ethnic background.<br>• Our non-English-speaking population has increased; in fact, it has doubled in five years to about 3.5% of the school population.<br>• We had an increase in free and reduced-cost lunch program population in five years (from 41% to 48%), but are still short of receiving Title I funds. We know there are students who qualify whose parents have not completed the forms.<br>• Attendance is low: 10% of students are absent daily.<br>• Attendance of special education and English Language Learner students is especially low at 9th and 10th grade levels (approximately 86% and 70%, respectively).<br>• We still have violent offenses that we would like to eliminate.<br>• 9th grade retention rate is a concern, as it has increased to 14%.<br>• Our dropout rate is 22%.<br>• Our graduation rate has dropped to well below the state target: 71% graduated on time (decreasing over the past five years); 38% special education and 67% of the free/reduced-cost lunch program students graduated. Graduation rates by ethnicities are about equal. More females (73.5%) graduated than males (69.3%).<br>• About 31% of our parents are college graduates, with 35% having some college experience, and 27% with high school diplomas.<br>• PVHS has three administrators (one principal and two assistant principals) who have been in their current positions for less than one year. |

*continued*

**FIGURE 7.1** *(continued)*

**Paradise View High School Demographic Data Analysis**

***Implications for the School Improvement Plan***

- We have highly qualified teachers. We have a strong core of teachers: a high number of licensed teachers, 5+ years at this school, with advanced degrees. We have to determine how to use our expertise to start turning our failing outcomes around.
- We must get proactive with helping parents complete free/reduced-cost lunch program applications. There is a high population of disadvantaged students who need extra supports.
- With increased English Language Learners (ELL) and free/reduced-cost lunch students, we need to review current services and how we can better serve our ELL and disadvantaged students.
- We have been resisting a full-blown Response to Intervention (RtI) system. We need to continue to develop and implement our RtI plan to address many of our concerns and improve the current strengths.
- A large number of students are going to two- or four-year colleges, but PVHS has a low graduation rate. Students are exiting this school because they are not making the grades or they are not enrolling in the school initially (even if in our school district) due to lack of specific programs offered. Let's talk with students and graduates to find out more about this phenomenon.
- We need to provide more support and monitoring at the 9th grade level.
- We need to look at how we can prevent violent offenses.
- Differences in administration may result in differences in decisions to suspend. It is time to set up a consistent behavior plan.
- We need to look at the extent of vaping as a new problem, as well as the effectiveness of consequences. Does the new law have an impact?

- *Green*. This process or program is important to our vision and everyone is implementing it as intended.
- *Yellow*. This process or program is important to our vision and not everyone is implementing it as intended.
- *Pink*. This process or program is optional or a duplication of efforts and needs to be tweaked to align to our vision.
- *Red*. This process or program is not important to our vision and should be eliminated.

Staff easily disposed of the red-coded processes and programs, there was a lively discussion about the pink- and yellow-coded processes and programs, but there were *no green-coded processes*. The biggest implication that resulted from their discussion was that the school did not have the vision to know which processes and programs needed to be implemented.

**FIGURE 7.2**

**Paradise View High School Perceptions Data Analysis**

| ***Strengths*** | ***Challenges*** |
|---|---|
| *Student questionnaire responses*: Overall, average responses were in agreement, with highest being the following:<br>• Doing well in school makes me feel good about myself.<br>• I feel ready for the real world with reference to my ability to learn on my own outside of the classroom.<br>• I feel ready for the real world with reference to my ability to read.<br>• My teachers expect me to do my best.<br>• My teachers expect all students to do their best.<br>• I am doing my best in school.<br>• In my classes, time is spent listening to the teacher talk.<br>Disaggregated responses showed no real differences for students by gender, ethnicity, plans after graduation, and grade level (although seniors were a little more positive).<br>With respect to the grade level they were when entered the school, those who entered as sophomores were generally the most positive.<br>Students' open-ended responses about the strengths of the school focused on their friends and electives.<br>*Staff questionnaire responses:* Most overall averages were in agreement. Those in strong agreement were:<br>• I love seeing the results of work with students.<br>• I believe learning can be fun.<br>• It is important to communicate often with parents.<br>• I believe student achievement can increase through: | *Student questionnaire responses*: None of the overall average responses were in strong agreement. The lowest overall averages (in disagreement) were:<br>• I find what I learn in school to be relevant to real life.<br>• I feel that I am in charge of what I learn.<br>• This school is fun.<br>• My teachers know me well.<br>• My teachers make learning fun.<br>• In my classes, time is spent doing work I find meaningful.<br>• My teachers are understanding when students have personal problems.<br>Disaggregated responses showed that students not affiliated with a club or extracurricular activities were in most disagreement with feeling like they belong at the school, that school is fun, that they find what they are learning in school to be relevant to real life, that school is preparing them for what they want to do after high school, and that they like this school.<br>Students affiliated with speech/debate do not feel the work they are asked to do at the school is challenging, that what they are learning in school is relevant to real life or meaningful, they do not like the students at the school, they do not think other students like them, they do not feel they belong at this school, and they neither like the school, nor do they think it is fun. These students also stated they do not have opportunities to choose their own projects, that they do not feel they are in charge of what they learn, and that they do not feel successful at school. They especially do not learn well when working in small groups.<br>The other disaggregation that stuck out was the grade level at which students entered the school. Students who entered the school as juniors were in disagreement with most questions.<br>Students' open-ended responses about what needs to improve at the school focused on teachers caring about them and the subjects they teach, teacher attitudes, and poor teaching. They want learning to be more fun and less boring. They want teachers to communicate with each other to make assignments make sense—together. They also want better food and a better schedule.<br>Staff questionnaire responses: The lowest overall averages (in disagreement) were:<br>• Morale is high on the part of teachers.<br>• Morale is high on the part of support staff. |

*continued*

**FIGURE 7.2** *(continued)*

**Paradise View High School Perceptions Data Analysis**

| ***Strengths*** | ***Challenges*** |
|---|---|
| — parental involvement<br>— providing a threat-free environment<br>— addressing learning style<br>— close personal relationships with teachers and students<br>• I believe every student can learn.<br>• I believe learning can be fun.<br>• I love to teach.<br>• I work effectively with ethnically and racially diverse students.<br>Support staff responses were in higher agreement than classroom teachers on every question.<br>Staff with four to six years of experience strongly agreed that they love seeing the results of their work with students, followed closely by staff with one to three years of experience. While in strong agreement, the more experience staff had, the lower their agreement with this item.<br>In open-ended responses, the most often written-in strength of the school was related to hard working, caring staff. | • This school has a good public image.<br>• Teachers in this school communicate with one another to make learning consistent across grade levels.<br>In addition to the items above, classroom teachers were in disagreement with the following items:<br>• Morale is high on the part of students.<br>Staff with 11 or more years of experience disagreed the most with morale being high on the part of teachers, support staff, students, and administrators.<br>Staff with four to six years of experience disagreed with the following:<br>• This school has a good public image.<br>• Quality work is expected of all adults in this school.<br>• I believe student achievement can increase through:<br>— Using ongoing student assessments related to state standards<br>— Teacher use of student achievement data<br>In open-ended responses, the most often written-in response to *what needs to be improved* in the school was related to morale, public image, school pride, consistency, caring relationship with one another and students, special education/inclusion, and better teachers. |

***Implications for the School Improvement Plan***

- We need to understand the correlation between student perceptions and academic achievement.
- Our current means of communication is unacceptable to students, staff, and parents. Some parents might not have access to email and our online grading system. We need to administer the parent questionnaire.
- All staff need to help one another improve attitudes and act like caring adults at all times.
- All staff have to really believe that all students can learn and that the implementation of our vision is our commitment to making that happen.
- We need to get honest and serious about improving the quality of teaching at this school.
- Our instruction has to become more real-life relevant, connected across subjects, and fun.
- We need to get serious about dropout and transfer retention. You can see why the students want to leave. A student focus group could help us get good ideas.
- We need to develop an ongoing public relations plan to improve our school reputation and public perception.

**FIGURE 7.3**

**Paradise View High School Student Learning Data Analysis**

### *Strengths*

Our small attempts at RtI are paying off:

Math Workshop students moved up an average of 3.02 grade levels in the first semester.

Math Workshop served 20 students, with four exiting after one semester (20%).

Reading Workshop students moved up an average of 1.325 grade levels in the first semester.

Reading Workshop served 15 students, with one exiting after one semester (7%).

### *Challenges*

With only one year of students taking the new state standards test, we can see we have a long way to go. We see all of our state results as challenges:

- English Language Arts: 46% Proficient
  - 59% females were proficient compared to 38% males
  - 9% of special education were proficient compared to 52% of students not in special education
  - 42% of low SES students were proficient compared to 53% not low SES
- Mathematics: 36% Proficient
  - There was about an equal percentage of females and males proficient
  - 9% of low SES students were proficient compared to 23% not low SES
  - (Special education not reported)

**ACT Results Are Shown and Discussed Below**

**ACT Results: Last Year**

| | Below Benchmark | | | At Benchmark | | | Above Benchmark | | |
|---|---|---|---|---|---|---|---|---|---|
| | 9th | 10th | 11th | 9th | 10th | 11th | 9th | 10th | 11th |
| English | 43% | 48% | 61% | 11% | 6% | 3% | 46% | 46% | 36% |
| Mathematics | 79% | 76% | 86% | 3% | 6% | 4% | 18% | 18% | 10% |
| Reading | 72% | 71% | 79% | 4% | 7% | 2% | 24% | 23% | 19% |
| Science | 79% | 76% | 89% | 3% | 8% | 3% | 18% | 17% | 8% |

- *English:* Just under half of the 9th and 10th graders received scores above benchmark and below benchmark; almost two-thirds of 11th graders were below benchmark.
- *Mathematics:* No more than 18% were above benchmark; a majority of students in grades 9–11 were below benchmark.
- *Reading:* Less than one quarter of our 9th through 11th graders were above benchmark; a majority of 9th through 11th graders were below benchmark.
- *Science:* Approximately 18% of 9th and 10th graders were above benchmark; a large majority of 9th through 11th graders were below benchmark.
- Students at benchmark represented the smallest groups in general, in all subject areas.

*continued*

**FIGURE 7.3** *(continued)*

**Paradise View High School Student Learning Data Analysis**

***Implications for the School Improvement Plan***

- Classroom instruction is not differentiated to meet diverse needs (i.e., disadvantaged, English Language Learners, special education, diversity).
- There is a lack of professional learning in the area of providing instruction to disadvantaged, students with learning disabilities, diverse students, and English Language Learners (ELL).
- There is a lack of professional learning in differentiated/formative instruction in recent years.
- Formative instruction using assessment data is not consistently implemented schoolwide.
- 9th grade core academic area teachers lack stability from year to year. Our strongest teachers should be placed in the 9th grade core classes.
- 9th graders have difficulty transitioning to rigorous level of high school work and the amount of homework required. We must revamp 9th grade services and requirements.
- School has not consistently targeted how to improve the Special Education inclusion program.
- Instruction schoolwide lacks consistency.
- The longer students are at the school, the lower their scores.
- Our targeted focus in the Reading and Mathematics Workshops appears to be having positive impacts. Let's build on this success. We need to keep going with our RtI plan.
- We have a large Special Education group with high needs that affects special education and the overall graduation rate.
- We need to analyze if a rigorous Senior Project for all students is impacting completion of English or Social Studies credits or are students dropping out to pursue a GED during their senior year.
- Special Education transitions from resource room settings to full inclusion are not working.
- Math curriculum has changed from an integrated math curriculum to a traditional math curriculum, and then to the state-generated curriculum—all are failing.
- Lack of a stable, cohesive math department.
- While reading proficiency met the state target, it is lower than expected and in comparison to other state high schools.
- Instruction in all classes transitioning to the new standards has not prepared students for the new state standards assessment.

How can we improve processes and programs to achieve the vision for our students if we don't have a vision? Excellent point. Figure 7.4 shows a portion of the final analysis of the school processes data, all of which needed further work.

**FIGURE 7.4**

**Paradise View High School Processes Analysis: Implications for the School Improvement Plan**

**Instructional Processes**

*Differentiated Instruction*

Staff clearly do not understand differentiated instruction. We need professional development to clarify what differentiated instruction looks like in the classroom, in every subject, and with integrated subjects. Teachers need time to plan—together—for differentiation of instruction. Classrooms need to be monitored to ensure implementation and to provide suggestions for improvement on an ongoing basis.

*Literacy Across the Curriculum*

Teachers need to continue their schoolwide focus on literacy across the curriculum. Expectations need to be clarified for implementing literacy strategies, and we need to revisit common strategies and themes schoolwide. Teachers would benefit from professional development and strategy suggestions specific to content area. Teachers would benefit from collaboration time with other departments.

*Mathematical Practices Across the Curriculum*

Teachers need to continue their schoolwide focus on mathematical practices across the curriculum. Not all teachers understand how to implement mathematical practices in their own content area. Expectations need to be clarified. Teachers need professional development and strategy suggestions specific to content area. Teachers would benefit from collaboration time with other departments, especially the math department. How can teachers use mathematics to differentiate instruction? The math department needs to provide guidance on moving mathematical practices across the curriculum.

*Schoolwide Standards Implementation*

We do not know how well we are implementing the standards in every classroom. Teachers would benefit from having more examples of how to align standards in their content areas. Teachers would benefit from having more reading and writing strategies that support implementation of the standards. Professional development, collaboration with implementation, coaching, monitoring, and evaluation are needed.

*Depth of Knowledge*

Professional development is needed to move instruction and learning from DOK 1–2 to DOK 3–4. Teachers need to focus on this in collaborative groups.

*Cycle of Instruction*

Instructional coaches need to provide coaching and feedback to strengthen each teacher's cycle of instruction. Cycle of instruction needs to be discussed and supported in collaborative groups, department meetings, and staff meetings.

*Curriculum Mapping*

Not all teachers have been on board with curriculum mapping. It is not an option and needs to be done by all teachers to include all aspects of instruction. Instructional scaffolding needs to be clear in every subject area and aligned across grade levels.

*Examining Student Work*

Staff would like to begin analyzing student work as a means of sharing how to integrate literacy and math across the curriculum, to make instruction related to real life, and more project-based. We need to create a plan and schedule for how this can happen throughout the year. One focus should be to help students scaffold project construction to their senior projects.

*Remedial Instruction*

We need to change our thinking about retention and remediation. A small group needs to create a plan to ensure every student is learning and excelling at grade level.

*continued*

**FIGURE 7.4** *(continued)*

**Paradise View High School Processes Analysis: Implications for the School Improvement Plan**

**Programs**

*Inclusion—Special Education*

The way we are doing inclusion is not working. We need professional development to clarify instructional expectations for all staff members and consider alternatives. The school needs to continue to monitor data relating to inclusion (e.g., failure rates, graduation rates). Can RtI replace this model?

*AVID—Advancement via Individual Determination*

We need to continue schoolwide implementation of AVID instructional strategies training. All staff need to know the purpose, expectations, benefits, strategies, results, and program requirements. Implementation needs to be monitored and evaluated for fidelity and to measure results.

*Credit Recovery*

What do the data tell us about credit recovery? Who are the students, what courses are being failed, and what are graduation rates? Why are students failing courses required for graduation? Does our program get them back on track?

*Advanced Placement*

We need to agree across subject areas what the purpose is and how we can support all students enrolled in AP courses. We need to increase the number of students taking and passing the AP examinations.

*9th Grade Transition*

The school needs to look at the needs of 9th grade students to help them succeed (academically and socially). The school needs to continue to develop the Transition to High School curriculum. We need to put some of our best teachers here to prevent failure.

*Senior Project*

Senior project needs to include all students. The projects should be multifaceted, researched, community based, and real world related. A small group needs to put together clear requirements and share with staff and students. Projects need to be required in all grade levels to build up to their Senior Projects.

*Workplace Readiness*

Expand to serve more students. Get all kids ready for the "real world." We need to make it help us implement standards—not be a detriment to standards implementation.

**Organizational Processes**

*Professional Learning Communities*

Professional Learning Community time has too many uses (i.e., data teams, professional development, school improvement). We need to dedicate the time for data analysis and instructional improvement, the way it was intended. How Professional Learning Communities should work needs to be clarified.

*Communication*

School needs to develop an effective system for supporting teachers who demonstrate marginal performance in instructional practice.

## Reviewing the Four Types of Data

The staff analyzed two types of data in two extended staff meetings—demographics and perceptions at one meeting and student learning and school processes in another. Their third staff meeting was devoted to aggregating all the data, or looking across the strengths, challenges, and implications of the different types of data to determine what had to go into their continuous school improvement plan.

The leadership team had earlier created a document of the strengths, challenges, and implications for the school improvement plan for the four types of data—demographics, perceptions, student learning, and school processes—and made copies for each staff member. Each staff member made her or his own analysis of the data results. Staff lined up the implications for the four types of data, lassoed common ideas, and compared their thinking with their small-group colleagues, which then was compared with other small groups' analyses. Figure 7.5 presents a side-by-side view of the implications of the four types of data. After lassoing common ideas, it soon became obvious what had to go into their continuous school improvement plan. The final analysis is shown in table form in Figure 7.6.

## Contributing Cause Analyses

Pleasant Valley High School staff used the problem-solving cycle to analyze what staff agreed were their biggest problems or challenges. They needed to get to the real contributing causes, using data, as opposed to only their assumptions. They wanted to tackle the first problem together, which was their low graduation rate. Figure 7.7 presents the 20 brainstormed hunches that staff came up with as to why the problem existed. (It was a great way to clarify staff thoughts about the problem.)

Staff next followed up with questions and data needed to answer the questions to know more about the problem. Figure 7.8 shows the questions they needed to answer with data to know more about the root of the problem and how to create a real solution to the real problem (as opposed to a solution to a symptom). Staff were anxious to know what the data would tell them.

**FIGURE 7.5**

**Side-by-Side View of Paradise View High School Data Implications**

| *Demographics* | *Perceptions* | *Student Learning* | *School Processes (outlined in detail in Figure 7.4)* |
|---|---|---|---|
| • We have highly qualified teachers. We have a strong core of teachers: a high number of licensed teachers, five-plus years at this school, with advanced degrees. We have to determine how to use our expertise to start turning our failing outcomes around.<br>• We must get proactive with helping parents complete free/reduced-cost lunch program applications. There is a high population of disadvantaged students who need extra support.<br>• With increased English Language Learners (ELL) and free/reduced-cost lunch students, we need to review current services and how we can better serve our ELL and disadvantaged students.<br>• We have been resisting a full-blown Response to Intervention (RtI) system. We need to continue to develop and implement our RtI plan to address many of our concerns and improve the current strengths. | • We need to understand the correlation between student perceptions and academic achievement.<br>• Our current means of communication are unacceptable to students, staff, and parents. Some parents might not have access to email and our online grading system. We need to administer the parent questionnaire.<br>• All staff need to help one another improve attitudes and act like caring adults, at all times.<br>• All staff have to really believe that all students can learn and that the implementation of our vision is our commitment to making that happen.<br>• We need to get honest and serious about improving the quality of teaching at this school.<br>• Our instruction has to become more real-life relevant, connected across subjects, and fun.<br>• We need to get serious about dropout and transfer retention. You can see why the students want to leave. A student focus group could help us get good ideas. | • Classroom instruction is not differentiated to meet diverse needs (i.e., disadvantaged, English Language Learners, Special Education, diversity).<br>• There is a lack of professional learning in the area of providing instruction to disadvantaged, students with learning disabilities, diverse students, and English Language Learners (ELL).<br>• There is a lack of professional learning in differentiated/formative instruction in recent years.<br>• Formative instruction using assessment data is not consistently implemented schoolwide.<br>• 9th grade core academic area teachers lack stability from year to year. Our strongest teachers should be placed in the 9th grade core classes.<br>• 9th graders have difficulty transitioning to rigorous levels of high school work and the amount of homework required. We must revamp 9th grade services and requirements.<br>• School has not consistently targeted how to improve the Special Education inclusion program.<br>• Instruction schoolwide lacks consistency. | **Instructional Processes**<br>Differentiated Instruction<br>Literacy Across the Curriculum<br>Mathematical Practices Across the Curriculum<br>Schoolwide Standards Implementation<br>Depth of Knowledge<br>Cycle of Instruction<br>Curriculum Mapping<br>Examining Student Work<br>Remedial Instruction<br>**Programs**<br>Inclusion—Special Education<br>AVID—Advanced via Individual Determination<br>Credit Recovery<br>Advanced Placement<br>9th Grade Transition<br>Senior Project<br>Workplace Readiness |

| ***Demographics*** | ***Perceptions*** | ***Student Learning*** | ***School Processes (outlined in detail in Figure 7.4)*** |
|---|---|---|---|
| • A large number of students are going to two- or four-year colleges, but PVHS has a low graduation rate. Students are exiting this school because they are not making the grades or they are not enrolling in the school initially (even if in our school district) due to lack of specific programs offered. Let's talk with students and graduates to find out more about this phenomenon.<br>• We need to provide more support and monitoring at the 9th grade level.<br>• We need to look at how we can prevent violent offenses.<br>• Differences in administration may result in differences in decisions to suspend. It is time to set up a consistent behavior plan.<br>• We need to look at the extent of vaping as a new problem, as well as the effectiveness of consequences. Does the new law have an impact? | • We need to develop an ongoing public relations plan to improve our school reputation and positive public perception. | • The longer students are at the school, the lower their scores.<br>• Our targeted focus in the Reading and Mathematics Workshops appears to be having positive impacts. Let's build on this success. We need to keep going with our RtI plan.<br>• We have a large Special Education group with high needs that affect special education and the overall graduation rate.<br>• We need to analyze if a rigorous Senior Project for all students is impacting completion of English or Social Studies credits or are students dropping out to pursue a GED during their senior year.<br>• Special Education transitions from resource room settings to full inclusion are not working.<br>• Math curriculum has changed from an integrated math curriculum to a traditional math curriculum, and then to the state generated curriculum—all are failing.<br>• Lack of a stable, cohesive math department.<br>• While reading proficiency met the state target, it is lower than expected and in comparison to other state high schools.<br>• Instruction in all classes transitioning to the new standards has not prepared students for the new state standards assessment. | **Organizational Processes**<br>Professional Learning Communities<br>Communication |

**FIGURE 7.6**

**Analysis of Paradise View High School Data Implications**

| *Instruction* | *Assessment* | *Curriculum* | *Climate* | *Data* | *Professional Learning* |
|---|---|---|---|---|---|
| • Teachers need to improve their instruction, set goals, and be willing to be coached for improvement.<br>• Every class should be engaging, real-life relevant, and rigorous.<br>• Instruction should build on the previous years' learning.<br>• Instruction should be integrated and project-based.<br>• We need to move solid teachers to the 9th grade and improve services for the incoming students.<br>• We have to determine how to improve reading everywhere.<br>• Instruction needs to incorporate DOK levels 3 and 4. | • We need to start the year with assessments to know where our students are and to adjust instruction to the students.<br>• We need diagnostics to assist us with our understandings of how to reach every student.<br>• Common formative assessments need to be agreed upon in subject area and across grade-level groups to align to the standards.<br>• As often as possible, assessments should be project-based, and a part of the work students do every day.<br>• We need to have regular examinations of student work to align curriculum and to improve instruction for every student. | • All curriculum must be aligned to standards, and aligned across grade levels.<br>• The math department needs to commit to a curriculum and improvement.<br>• All subjects should be mapped and scaffolded before the beginning of next year.<br>• We need to begin integrating subjects.<br>• We need to implement a character education curriculum congruent with our new behavior plan.<br>• We need a plan and implementation strategy for college and career preparation. | • We need a dropout prevention program that is informed by data analysis and focus groups.<br>• We need a clear behavior plan, complete with rituals and routines, that is implemented with consistency in every corner of the school.<br>• Every teacher needs to build positive, caring relationships with students.<br>• All staff need to act congruent with beliefs that all students can learn.<br>• We need to welcome every student to school, every day. | • We need to have parents complete the free/reduced-cost lunch forms the first time they register their child. They will remain qualified until the parents tell us otherwise.<br>• Continue the schoolwide data analysis work to understand the relationship of student perceptions and academic achievement.<br>• Continue the data analysis work to ensure we are meeting the needs of all students.<br>• Continue to do the data analysis work to understand more about our dropouts and chronically absent students. | • *Improve instruction* to meet the needs of our students, especially disadvantaged, students with special needs, diversity, non-English speakers, and those needing remediation.<br>• *RtI system*, specifically how to improve core instruction and interventions to meet the needs of *every* student.<br>• *Differentiated instruction* that incorporates literacy and mathematics across the curriculum.<br>• *Depth of Knowledge* —we all need to learn how to ask better questions that draw out student understanding. |

| *Collaboration* | *Leadership* | *Partnerships* | *Senior Projects* | *Vision/Plan* | *RTI/Special Education* |
|---|---|---|---|---|---|
| • Teachers need to collaborate to improve the cycle of instruction, and to ensure every student is learning.<br>• Need to focus on student work, student data, DOK 3 and 4, and high-quality instruction.<br>• Teachers need to ensure implementing literacy and math across the curriculum. | • We want our new leaders to lead us to a new shared vision that will help us attain our desired outcomes.<br>• We need a new communication structure for teachers, administrators, students, and parents.<br>• Classrooms need to be monitored for quality and implementation of the shared vision. | • We need to build a strong partnership with the middle school to help students transition to high school and prevent potential dropouts.<br>• We need to build a strong partnership with the community college, so students can get joint credits.<br>• We need to develop an ongoing public relations plan to improve our school reputation. | • Senior projects need to be redefined to be a culmination of high school experience, with a community focus.<br>• Every grade level needs to require projects that scaffold to the senior project and that integrates instruction. | • We need to create a shared vision, and a concrete plan to implement the vision.<br>• Our vision needs to be implemented, monitored, and evaluated according to our agreements in the Program Evaluation Tool. | • RtI needs to be designed, implemented, monitored, and evaluated throughout our school, as spelled out in the RtI Program Evaluation Tool and the RtI Implementation Guide created by the RtI Team.<br>• *Special education*—the Program Evaluation Tool needs to be used to improve services, especially in consideration of the RtI system. |

**FIGURE 7.7**

**Paradise View High School Twenty Hunches as to Why Graduation Rate So Low**

1. Students don't care.
2. Students don't want to work or go to college.
3. Parents don't care.
4. Students find no value in working hard in school.
5. There are no consequences that they care about for not graduating.
6. Middle school does not prepare them for high school.
7. Students do not read.
8. Students are bad in math.
9. It is not cool to be smart here.
10. There are not enough extracurricular activities to keep students interested in this school.
11. Students only work hard in their electives.
12. Our culture does not encourage high school completion enough.
13. We are not preparing students for college and careers.
14. Courses are boring.
15. Courses are not relevant.
16. We don't care enough.
17. Our expectations are too low.
18. Our teaching is inconsistent and poor for the most part.
19. We do not know what students need or how to teach to their needs.
20. We do not work together for the benefit of the students.

The leadership team then assigned other challenges that came out of the data analysis work to each of the 10 small groups. The challenges included:

- Lack of student proficiency in mathematics
- Lack of student proficiency in English language arts
- RtI not working
- Inclusion not working
- Professional Learning Communities not working
- Special Education students not performing or attending school
- Student attendance
- Freshmen failures
- Students not engaged in learning
- Lack of high-quality instruction in every classroom

Each small group put its problem-solving cycle on chart paper. The staff did a gallery walk with markers in hand and added comments next to the other groups' work. A few things became clear—no matter the issue, staff had many of the same hunches. That meant a couple of things. First,

**FIGURE 7.8**

**Paradise View High School Questions and Data to Learn More About Low Graduation Rate**

| ***Questions*** | ***Data Needed*** |
|---|---|
| 1. Who are the students who are not graduating? | Dropout data, disaggregated by grade level, gender, ethnicity/race, special education, English learners, attendance, behavior, number of credits, course grades, state assessment results over time, middle school performance, and attendance. |
| 2. When are the students dropping out? | |
| 3. What are the predictors of dropping out? | |
| 4. Why are students not graduating? | Focus group discussions or phone calls with dropouts can help the school secure the answers to questions, 4, 5, 6, and 7.<br><br>Any additional data we can get on our former students can help us learn why they chose to leave school when they did. |
| 5. What courses can we offer to keep students engaged in school? | |
| 6. What can we do to prevent dropping out of school? | |
| 7. What happens to the dropouts? | |
| 8. What would it take to partner with the middle school to prevent dropouts, and with the community college to offer joint credit courses? | Set up a team to follow up the partnership discussion with the middle school and community college. |
| 9. What can we do to ensure every classroom is of highest quality? | After researching better ways to reach all students, we need to create a shared vision that has high-quality instruction as a focus. |

staff perceptions reflected the way they were acting; and, secondly, many of the same changes would eliminate most of the problems. After the data outlined in each problem-solving cycle were analyzed, staff added the implications to the other implications for the continuous school improvement plan, if they were not already listed in their implications. Most implications were already there, but maybe with a different tilt.

Each group was additionally tasked with the challenge of doing research on its "challenging" topic to determine how staff might think about the topic differently going into the creation of the shared vision. Time was allocated during the week for book studies and research. An RtI team was identified and assigned to attend state-sponsored RtI training to design a comprehensive RtI system for the school. Other teams attended available training when they could.

Each team was able to share what they had learned in staff meetings over the next couple of months. All staff were encouraged to continue studying new ways of reaching the students and to begin implementing the ideas when they could. The leadership team thought it would be useful to use a tool they saw in their continuous school improvement training called the Process Measurement Planning Table to describe what all the processes they were researching might look like, and the evidence that would indicate each is really being implemented. Figure 7.9 shows the Paradise View High School Process Measurement Planning Table with nine of the processes staff intended to implement.

## Creating a Shared Vision

Staff felt they had uncovered many challenges in their school. They believed it was time to design a school in which they were proud to say they were employed. Most of all, they wanted a school that would help *every* student excel. In a student-free professional learning day, staff committed to creating and implementing a shared vision. The principal began by saying that there would be one major guiding principle: any staff member who planned to work in this school had to *believe* and act like *every* student can, and will, learn. The vision they were about to create would have that mantra as its starting and ending point.

Fresh from their book studies and research about strategies to improve the school, staff were asked to list what they believed impacts learning for their students, with respect to curriculum, instruction, assessment, and the learning environment. In other words, what did they already know made a difference for their students, either through experience or from the research

**FIGURE 7.9**

**Paradise View High School Process Measurement Planning Table: Processes That Will Be Implemented**

| *What process do you want to measure?* | *What do you want the process to look like?* | *How can this process be measured?* |
|---|---|---|
| **Schoolwide Standards-Based Curriculum** | All curriculum is mapped and aligned to state standards, content, and grade-level expectations; a continuum of learning that makes sense to students; agreed upon across all grade levels. | • Curriculum mapping<br>• Classroom observations<br>• Student achievement results (student data and student work)<br>• Staff, student, parent, and standards questionnaires<br>• Instructional coherence |
| **Literacy Across the Curriculum** | The literacy skills—reading, writing, speaking, listening, viewing, presenting, and critical thinking—are evident in all content areas. | • Classroom observations<br>• Lesson plans<br>• Student achievement results (student data and student work)<br>• Collaborative groups<br>• Instructional coherence<br>• Program Evaluation Tool |
| **Mathematical Practices Across the Curriculum** | Skills such as interpreting graphs, identifying trends, analyzing information, and identifying patterns and relationships are evident in all content areas. | • Classroom observations<br>• Lesson plans<br>• Student achievement results (student data and student work)<br>• Collaborative groups<br>• Instructional coherence<br>• Program Evaluation Tool |
| **Instruction** | Agreed-upon strategies implemented in every classroom, including small- and large-group instruction, flexible groupings, differentiated instruction, scheduling; designed to meet the needs of *every* student. Cycle of Instruction is used and DOK 3–4 is used regularly. | • Measures of instructional coherence<br>• Classroom/teacher observations<br>• Student achievement results (student data and student work) builds from one year to the next<br>• Staff and student questionnaires<br>• Collaborative groups |

*continued*

**FIGURE 7.9** *(continued)*

**Paradise View High School Process Measurement Planning Table: Processes That Will Be Implemented**

| *What process do you want to measure?* | *What do you want the process to look like?* | *How can this process be measured?* |
|---|---|---|
| **Assessments** | Formative assessments aligned to the standards and grade-level expectations, progress monitors that measure responsiveness to interventions, and high-stakes summative assessments are used to know what students know and do not know. | • Assessment inventory<br>• Classroom/teacher observations<br>• Student achievement results (student data and student work)<br>• Staff and student questionnaires |
| **RtI** | Our instruction and assessment system seamlessly provides the learning forum that accelerates *each* student's learning. All teachers and staff work together to ensure *every* student's success. | • RtI Program Evaluation Tool<br>• RtI system flowchart<br>• Interventions that are accelerating each student's learning, while keeping students in their regular classes<br>• RtI assessment results<br>• Grades, student work |
| **Staff Collaboration** | Teachers meet in teaching teams to review student progress (student learning data and student work), to improve the implementation of the vision, and to adapt processes, as needed. | • Staff questionnaire<br>• Leadership structure<br>• Instructional improvement<br>• Student data<br>• Student work review |
| **Learning Environment** | Students come to school and feel like they belong, are challenged, and are cared for.<br>Teachers feel supported and that they are working in a collaborative environment; teachers have high expectations for students and believe every student can learn.<br>Parents feel welcome at the school, and know what they can do to support their child's learning; effective home-school communications. | • Student, staff, and parent questionnaires<br>• Behavior data<br>• Teacher and student attendance<br>• Dropout and graduation rates |
| **Leadership** | Leadership structure that helps everyone implement the vision; supportive of all staff, students, and parents; supports the continuous improvement of the organization and all personnel. | • Student, parent, and staff questionnaires<br>• Leadership structure that helps everyone implement the vision<br>• Evaluation tools and strategies |

recently conducted and shared. Individuals wrote down their thoughts. That thinking was merged in small groups, and the small group thinking was merged to form core values and beliefs for the whole staff, as shown in Figure 7.10.

After completing the core values and beliefs, staff were asked to come to consensus on the purpose of the school. After much deliberation, they agreed that the purpose of Paradise View High School was to prepare students to be anything they wanted to be in the future. From this purpose, staff came to agreement on a mission statement:

> Paradise View High School provides a rigorous, relevant, and standards-based curriculum in a welcoming, supportive learning environment that enlightens and engages all students to become life-long learners who are effective communicators and creative, collaborative, critical thinking, self-directed individuals.

The next question was, What had to be in curriculum, instruction, assessment, and the learning environment in order to achieve the mission, taking into account the core values and beliefs?

Staff began with an example vision, and adjusted it to incorporate their core values and beliefs. One quarter of the staff worked on curriculum, another quarter worked on instruction, assessment, or the learning environment. Each group presented information on its area. Some adjustments were made. Each group was given a different category to refine. After two rounds, staff felt very good about the resulting shared vision. All staff participated, and all staff committed to reviewing the vision after the leadership team had a chance to refine it. Staff also committed to implement the resulting vision with integrity and fidelity. Figure 7.11 presents the Paradise View High School shared vision for curriculum, instruction, assessment, and the learning environment.

To tie all the pieces together, to ensure implementation with integrity and fidelity, to establish monitoring, and to set up program evaluation, staff used the Program Evaluation Tool for their shared vision (see Figure 7.12). One more time, staff were divided into work groups to conquer the challenge. They brought the work together and refined it. At the end of the

**FIGURE 7.10**

**Paradise View High School Core Values and Beliefs**

| ***Curriculum***<br>*What we teach* | ***Instruction***<br>*How we teach the curriculum* | ***Assessment***<br>*How we assess learning* | ***Learning Environment***<br>*How each person treats every other person* |
|---|---|---|---|
| • Advanced Placement and Honors<br>• Begin with end in mind; what we want our students to master<br>• Character education<br>• Community-based<br>• Consistency<br>• Containing basic factual knowledge to produce well-rounded students<br>• Coping skills in life<br>• Critical thinking<br>• Integrated fine arts<br>• Interdisciplinary<br>• Knowledge and skills for college and career<br>• Multicultural and multiethnic<br>• Respect, rigor, relevance, and responsibility—no matter the subject<br>• Oral communication<br>• Progression of skills and concepts from grade level to grade level<br>• Relevance for self, community, and global society<br>• Standards-based<br>• Varied electives | • Academic conversation skills<br>• Allow students to explore and engage<br>• Challenging<br>• Check for understanding and reteach when necessary<br>• Connections to prior learning<br>• Differentiated<br>• Direct instruction—kept to a minimum<br>• Engaging<br>• Follow logical steps in the learning progression<br>• Group work that functions and models real life<br>• Hands-on activities<br>• In ways that lead students to think independently<br>• Incorporate technology<br>• Lessons that engage critical thinking and application of knowledge<br>• Literacy and numeracy in all subjects<br>• Multiple intelligences, learning styles, and interests | • Alternative assessments<br>• Assess with as many means as possible to reach as many learning styles as possible<br>• Consider the individual goals of each student<br>• Consistent across all grade levels and subject areas<br>• Feedback and reflection<br>• Lab practicums<br>• Mindful of other classes in frequency<br>• More open-ended/less multiple choice<br>• Observation<br>• Performance-based<br>• Presentations<br>• Reward students' knowledge by allowing them to test out of courses and receive credit<br>• Rubrics<br>• Service learning projects<br>• Student awareness of quality work<br>• Summative assessments | • Accepting<br>• All adults on campus need to work to support each other to make the focus of this campus to be about *every* student learning<br>• Always better to light a candle than to curse the darkness<br>• Classrooms should be welcoming, clean, safe, organized, inviting, and orderly (set routines)<br>• Clear communication on all parts<br>• Compassion<br>• Consistency<br>• Culturally aware and nonbiased<br>• Develop relationships<br>• Drug-free environment<br>• Encouraging and caring<br>• Everyone held accountable<br>• Fair, fun, helpful, honest, inclusive, kind<br>• Instill maturity<br>• Nonviolent expression<br>• Obey the law<br>• Refer to the golden rule: treat others like you want to be treated |

| ***Curriculum***<br>*What we teach* | ***Instruction***<br>*How we teach the curriculum* | ***Assessment***<br>*How we assess learning* | ***Learning Environment***<br>*How each person treats every other person* |
|---|---|---|---|
| • Work ethic<br>• World languages | • Project-based<br>• Real-world connections<br>• Teach by example<br>• Teach so that all students are able to learn<br>• Teach to standards<br>• Teachers as facilitators; students as workers<br>• Work in partnership with our community | • Use of pre-assessments<br>• Written reflections | • Routines and expectations clearly stated<br>• Safe—physically and emotionally<br>• Socially appropriate behavior<br>• Teacher-student relationships form the basis in which learning takes place<br>• Teachers are approachable so students feel comfortable asking for help<br>• When students feel connected to school, they want to be here<br>• When students have supportive and caring relationships in school, they are more successful (peers, teachers, support staff) |

FIGURE 7.11

**Paradise View High School Vision: Curriculum, Instruction, Assessment, and Learning Environment**

*Paradise View High School provides a rigorous, relevant, and standards-based curriculum in a welcoming, supportive learning environment that enlightens and engages all students to become life-long learners who are effective communicators and creative, collaborative, critical-thinking, self-directed individuals.*

**Paradise View High School Vision for Curriculum**

| *Components* | *What It Will Look Like* | *Actions* | *Evidence* |
|---|---|---|---|
| **Curriculum is rigorous, relevant, and standards-based.** | Curriculum is aligned to:<br>• State standards<br>• Career and technical education programs of study | Teachers implement curriculum aligned to their standards.<br>Teachers plan instruction of their curricula that addresses:<br>• State standards<br>• Advanced Placement classes/Honors classes<br>• Interdisciplinary units<br>• Transition to high school<br>• Career and technical education programs of study<br>• Depth of Knowledge 3 and 4 activities<br>• Lesson plans incorporate use of technology | • Curriculum maps<br>• Assessment results<br>• Teacher lesson plans<br>• Syllabi<br>• State-approved curriculum<br>• Project- and problem-based lessons<br>• Walk-throughs<br>• Project-based assessments<br>• Senior projects |
| **The curriculum is horizontally and vertically aligned to the standards.** | Curriculum maps are aligned to the standards (horizontal alignment).<br>Departmental progression of standards reflects a continuum of learning for students (vertical alignment). | Teachers collaborate departmentally to align curriculum horizontally and vertically.<br>Collaboration with middle schools and the community college to ensure 6–12 progression of skills and knowledge to be college- and career-ready. | • Student work<br>• Teacher collaboration (planning days, looking at student work, student data)<br>• Observations<br>• 6–12 + progression of skills and knowledge to be college and career ready |
| **Literacy and mathematical strategies are integrated in all content areas.** | • Literacy and math standards are integrated in all courses.<br>• At least 70% of texts that students read should be informational text. The remaining 30% should be literary text. | • Teachers will receive professional development on integrating literacy and mathematical standards into content areas.<br>• Teachers will work in collaborative teams to support implementation. | • Professional learning<br>• Collaborative team work<br>• Student data<br>• Student work<br>• Senior projects<br>• Teacher lesson plans |

| Paradise View High School Vision for Instruction | | | |
|---|---|---|---|
| ***Components*** | ***What It Will Look Like*** | ***Actions*** | ***Evidence*** |
| **Instruction is aligned to the state standards and incorporates literacy and math strategies.** | • Instructional coherence is in place through horizontal and vertical alignment.<br>• Literacy and math strategies are incorporated across the curriculum.<br>• Instruction is reviewed and updated on a regular basis.<br>• Instruction aligns with standards and assessments.<br>• Students understand the standards that they are learning.<br>• Teachers understand the vertical progression of standards. | • Professional learning focuses on best practices for each content area, including understanding assessments and their uses.<br>• Professional learning provides instruction on integrating literacy and math practices in the classroom.<br>• Time is dedicated for department collaboration.<br>• Teachers are given time for vertical and horizontal collaboration.<br>• Standards are broken down for students to understand the end goal.<br>• Teachers plan for vertical articulation. | • Curriculum mapping<br>• Lesson plans<br>• Standards alignment is evident in student work<br>• Classroom observations<br>• Walk-throughs<br>• Collaborative team meeting minutes |
| **Instruction is differentiated to address needs of students.** | • Instruction is targeted to meet the needs of all students.<br>• Various learning styles are addressed through instruction.<br>• Resources, teaching and learning strategies, and assessments are differentiated to meet student needs.<br>• Response to Intervention (RtI) is in place and operates seamlessly.<br>• Modification and accommodations are effectively integrated to meet students' needs. | • Formative assessments are used to identify the learning needs of students and effectively drive instruction and interventions.<br>• Students complete learning-style inventories and teachers use the information to drive instruction.<br>• RtI is defined and integrated into instruction.<br>• Teachers create lessons designed to meet student learning styles.<br>• Teachers meet to analyze student work, identify individual student needs, and plan interventions.<br>• Teachers scaffold lessons to support all students' learning (including high-needs and high-achieving populations.)<br>• Teachers actively plan, implement, and reflect on their instructional strategies. | • Collaborative team minutes<br>• Student work<br>• Student interest inventory<br>• Student data<br>• Professional learning<br>• Walk-throughs<br>• Observations |

*continued*

**FIGURE 7.11** *(continued)*

**Paradise View High School Vision: Curriculum, Instruction, Assessment, and Learning Environment**

| *Components* | *What It Will Look Like* | *Actions* | *Evidence* |
|---|---|---|---|
| | **Paradise View High School Vision for Instruction** *continued* | | |
| **Instruction is rigorous and relevant.** | • Instructional tasks, questions, and assessments are developed at a DOK level 3 or 4 to develop creativity, innovation, and critical thinking skills.<br>• Text ratio is 30% literary and 70% informational texts.<br>• Teachers provide students with the opportunity to make connections and engage in project-based learning.<br>• Instruction is bell to bell.<br>• Consistent tracking of student performance to identify areas of strengths and weaknesses. | • Teachers use a variety of instructional strategies to promote and teach all academic vocabulary. Teachers provide opportunity for students to apply knowledge in curricular and extra-curricular activities, such as senior projects, and integrated subject projects.<br>• There is an explicit link between assessments and instruction to ensure rigor and relevance.<br>• Teachers implement common formative assessments to track student achievement. | • Student work<br>• Student projects<br>• ACT and other assessments<br>• Number of AP classes offered<br>• Number of students in AP courses<br>• Number of students taking the AP exams<br>• Number of students passing the AP exams |

| Paradise View High School Vision for Assessment | | | |
|---|---|---|---|
| ***Components*** | ***What It Will Look Like*** | ***Actions*** | ***Evidence*** |
| **Different types of assessments are used to measure student learning and inform instruction, and all are aligned to the State Standards.** | Assessment plans include:<br>• Multiple assessments to determine student progress,<br>• A variety of common course formative and summative assessments,<br>• Assessments that reflect depth of knowledge (DOK) levels 3–4,<br>• Authentic assessments that require students to apply their thinking such as performance-based or project-based assessments, and<br>• Real-world applications of knowledge. | All teachers know:<br>• What is being assessed in their subject.<br>• What standards are being assessed at the level below and the level above their course (vertical alignment) and how the students are performing.<br>Teachers collaborate and create common formative assessments.<br>Teachers use assessment date to drive instruction. Teachers collaborate on creating:<br>• Common formative assessments.<br>• A variety of opportunities for students to demonstrate knowledge and skills. | • Monitor in collaborative teams<br>• End of course exam<br>• Common formative assessments<br>• Standardized tests<br>• Curriculum-based assessments<br>• Student work<br>— Journal reflections<br>— Performance<br>— Class discussions<br>— Reflections<br>• Student-based assessments<br>• Performance assessment<br>• Observations<br>• Products<br>• Student Self-Assessment<br>• Competitions and extracurricular activities |
| **Students use assessment data to reflect upon and drive their own learning.** | • Teachers provide students opportunities to monitor and reflect on their progress towards standards attainment.<br>• Teachers provide specific, descriptive feedback in a timely manner to give students opportunities to monitor their own progress and processes. | • Students use assessment results to reflect and set goals.<br>• Students reflect on performance progress and set new goals. | • Student portfolios<br>• Student work<br>• Students use all assessment results |

*continued*

**FIGURE 7.11** *(continued)*

**Paradise View High School Vision: Curriculum, Instruction, Assessment, and Learning Environment**

| Paradise View High School Vision for the Learning Environment | | | |
|---|---|---|---|
| ***Components*** | ***What It Will Look Like*** | ***Actions*** | ***Evidence*** |
| **All students and staff work towards building positive relationships, with each other and the community.** | • Everyone treats one another with mutual respect.<br>• Everyone works through conflicts to find the best possible solutions.<br>• Peers engage in critical conversations focused on improving relationships.<br>• Staff model, teach, and reinforce expected behaviors.<br>• There is fidelity in implementation of expectations across all contexts.<br>• Everyone is actively involved in establishing community partnerships. | • Everyone will know how to interact appropriately with each other and be effective in accomplishing all tasks.<br>• Peers mediate disputes in a solution-focused manner.<br>• Everyone practices positive intent (reserves judgments).<br>• Faculty and students reach out to the community to build positive relationships.<br>• Everyone teaches and reinforces expected behaviors. | • Perceptual data<br>• Community partnerships<br>• School pride promoted by student activities (calendar)<br>• Reduction in school referrals |
| **All stakeholders collaborate constructively for continuous school improvement.** | • Problem-solving process is used to address concerns.<br>• Team agreements are clearly defined and upheld by all.<br>• Everyone works together, collaboratively, cooperatively, and shares knowledge resources, and skills across the curriculum.<br>• Staff use time effectively to advance school improvement initiatives.<br>• Everyone communicates respectfully in all modes (i.e., e-mails, in-person discussions, written correspondence).<br>• Everyone practices open and positive communication to all stakeholders (parents, community, students, colleagues).<br>• Leadership team meets regularly to discuss schoolwide goals.<br>• Communication flowchart is established and adhered to. | • Create online resource to house lessons accessible to all staff.<br>• Establish and clearly define the problem solving process.<br>• Ensure that resources to support implementation are available.<br>• Ensure implementation of Communication Flowchart.<br>• Clearly define expectation of communication behavior. | • Website<br>• Team norm posters<br>• Self-assessment<br>• Professional Learning<br>• Leadership Teams<br>• Regular use of flowcharts |

day, it became the leadership team's task to take the day's work and refine it. They were also tasked with drafting a process flowchart that reflected the shared vision that they would bring back to staff to review (see Figure 7.13). As next steps, each small group agreed to complete a Program Evaluation Tool for the programs and processes listed in Figure 7.9 and the others were assigned to research, clarify implementation, and set up each program or process for evaluation and implementation monitoring.

## Evaluating the Shared Vision

The data to evaluate how the school's programs and processes came together to carry out the shared vision were set up in the school's Logic Model, presented in Figure 7.14. Staff were excited to see the school's Theory of Change described as a comprehensive data analysis picture.

This analysis and planning work took most of the first year. One thing the staff saw as necessary was to lengthen and adjust their school day schedule so they could work with students who needed assistance, provide acceleration for students ready for more, and provide for more staff collaboration time. They added minutes to four days to create a special class period every week for every student to work on subjects in which they needed assistance, or acceleration. Each student was in charge of her or his personal goals. Time was created and dedicated for teacher collaboration as well.

As the second year began, staff were set for the new demographic data and ready to adjust the vision, programs, and processes on an ongoing basis throughout the year. As the new school year commenced, staff knew what they needed to do. They analyzed their data and end-of-year results, adjusted their processes, and recommitted to shared vision implementation. Staff were excited because they could see the benefit of their hard work; they knew they were now nimble enough to adjust, as needed, and most of all, they knew they were reaching most, if not all, students.

In an early year staff meeting, a few math teachers complained that the way they were "doing" RtI was too labor intensive for teachers. They wanted to go back to "pulling students out" of classes or to use special education

**FIGURE 7.12**

**Paradise View High School Shared Vision**

| **Needs Assessment** | **Purpose, Participants, and Outcomes** | | **Implementation** | | **Results** | |
|---|---|---|---|---|---|---|
| *What are the data showing about the need for the program or process?* | *What is the purpose of the program or process?* | *What are the intended outcomes?* | *How should the program/process be implemented with integrity and fidelity to ensure attainment of intended outcomes?* | *How is implementation being monitored?* | *How will results be measured?* | *What are the results?* |
| • We have 1,040 students, reflecting three major ethnicities.<br>• Fourteen percent special education—more than half from one ethnic population.<br>• Three-and-a-half percent of the student population are English Language Learners.<br>• Fewer than half of our students qualify for free/reduced-cost lunch; however, we are sure more could qualify. Some families do not sign up.<br>• Average daily attendance is 90.7% | Paradise View High School provides a rigorous, relevant, and standards-based curriculum in a welcoming, supportive learning environment that enlightens and engages all students to become life-long learners who are effective communicators and creative, collaborative, critical-thinking, self-directed individuals.<br><br>***Who is the program or process intended to serve?***<br><br>PVHS is intending to serve *every* student who comes through the doors. | When the purpose of PVHS is implemented as intended:<br>• Instructional coherence and a continuum of learning that makes sense for all students will be evident.<br>• What students learn in one grade level will build on what they learned in previous grade levels, and prepare them for the next grade level.<br>• Individual student achievement results will improve each year.<br>• All students will be proficient in all areas. No student will need to be retained. | • Before school starts, teachers in their subject area teams will develop curriculum and scaffolding maps, determine what they are going to teach, and the materials they are going to use across the three grade levels, so there is continuity and overlap in learning.<br>• These same teams meet regularly throughout the year to plan and adjust their instruction on the basis of student assessment results, and to share strategies for integrating literacy and mathematic strategies in their lessons. | • The Program Evaluation Tool is used to monitor the implementation of the major programs and processes.<br>• The vision implementation tool is used as a teacher self-assessment tool, coupled with an outside review by the curriculum coordinator and instructional coaches.<br><br>***How should implementation be monitored?*** | • Student, staff, and parent questionnaires will be administered to understand what the stakeholders are thinking and feeling about the school.<br>• *Student learning growth and proficiency will be measured through the state assessment, through common formative assessments, and progress monitoring tools. | • At the end of the first year of implementing our new vision, we saw more students becoming proficient, with more students showing at least a year's growth in learning.<br>• Our RtI system got started with some hiccups, but will be working well by next year.<br>• Student attitudes about school have improved in the past two years. Students are feeling challenged. |

| ***Needs Assessment*** | ***Purpose, Participants, and Outcomes*** | | ***Implementation*** | | ***Results*** | |
|---|---|---|---|---|---|---|
| *What are the data showing about the need for the program or process?* | *What is the purpose of the program or process?* | *What are the intended outcomes?* | *How should the program/process be implemented with integrity and fidelity to ensure attainment of intended outcomes?* | *How is implementation being monitored?* | *How will results be measured?* | *What are the results?* |
| • Attendance of special education and English Language Learner students is especially low at 9th and 10th grade levels (about 86% and 70%, respectively).<br>• Ten percent of students are absent each day.<br>• The percentage of students who are suspended has decreased (92.5% to 91.7%).<br>• 9th grade retention rate is a concern, as it has increased to 14%.<br>• Our dropout rate is 22%. | ***Who is being served? Who is not being served?***<br><br>The students who are proficient are being served. The many who are not proficient, dropping out, and transferring before graduation are not being served. | • Progress monitoring and common formative assessments conducted within the classroom will be utilized to identify struggling students and why they are struggling.<br>• Interventions matched to student needs will result in student learning increases for *every* student.<br>• All students at risk of low achievement are identified early and "failure" is prevented.<br>• Fewer students will be identified for special education. | • All students are screened within two weeks of school starting in their core classroom, and again at the middle and end of the year.<br>• The results of the screeners are reviewed by the RtI team to get a schoolwide view of the performance of all students, to set cut scores, and to allocate support.<br>• RtI team reviews the screening results with classroom teachers, and together they set up interventions for students at risk of failing. | We started monitoring as described above; however, we want to monitor more often next year.<br><br>***To what degree is the program being implemented with integrity and fidelity?***<br><br>In our first year, staff did a fabulous job of clarifying everything they were asked to implement. In our second year, we are hoping the implementation will become the way we naturally do business. | • Teacher effectiveness will be measured by walk-throughs by administration and implementation monitoring by peers.<br>• All major programs or processes will be monitored for implementation integrity and fidelity on a continuous basis to ensure the highest quality. | |

*continued*

**FIGURE 7.12** *(continued)*

**Paradise View High School Shared Vision**

| **Needs Assessment** | **Purpose, Participants, and Outcomes** | | **Implementation** | | **Results** | |
|---|---|---|---|---|---|---|
| *What are the data showing about the need for the program or process?* | *What is the purpose of the program or process?* | *What are the intended outcomes?* | *How should the program/process be implemented with integrity and fidelity to ensure attainment of intended outcomes?* | *How is implementation being monitored?* | *How will results be measured?* | *What are the results?* |
| • Our graduation rate has dropped to well below the state target; 71% graduated on time (decreasing over the past five years); 38% special education and 67% of the free/reduced-cost lunch program students graduated. Graduation rates by ethnicities are about equal. More females (73.5%) graduated than males (69.3%).<br>• Our students represent a high four-year college attendance compared to the state. | | • Students will not be placed in special education for the wrong reasons, such as teachers wanting students out of the classroom because of behavior or lack of learning response, poor test-taking skills, second language learning, or lack of adequate interventions.<br>• Referrals made for evaluation of special education are accurate. | • *All* students' progress is monitored according to curriculum assessments at the primary level, and weekly at the secondary and tertiary levels, with intervention progress monitoring tools.<br>• Teachers analyze assessment results in data teams and determine how to improve instruction to reach *all* students. | | | |

| Needs Assessment | Purpose, Participants, and Outcomes | | Implementation | | Results | |
|---|---|---|---|---|---|---|
| *What are the data showing about the need for the program or process?* | *What is the purpose of the program or process?* | *What are the intended outcomes?* | *How should the program/process be implemented with integrity and fidelity to ensure attainment of intended outcomes?* | *How is implementation being monitored?* | *How will results be measured?* | *What are the results?* |
| • Our 64 teachers have an average number of 10 to 12 years of teaching. Over half of our teachers have advanced degrees. Six are National Board Certified.<br>• More than 60% of our staff have been at our school for more than five years; teacher turnover has decreased.<br>• Ninety-five percent of teachers are licensed with over 90% of our classes being taught by Highly Qualified Teachers.<br>• Approximately 31% of our parents are college graduates, with 35% having some college experience, and 27% with high school diplomas. | | • Attendance and student engagement will improve because we are meeting the needs of every student.<br>• Behavior will improve because we are engaging all students in learning.<br>• There will be a decrease in dropouts, and the graduation rate will increase.<br>• Enrollment will increase because students will want to come here to learn.<br>• Decrease in dropouts.<br>• Higher graduation rate.<br>• Enrollment will increase because students want to come here to learn. | • If students are not making progress after **four to six data points**, the interventions may be changed, if the current interventions have been implemented with integrity and fidelity. RtI team will monitor.<br>• Instruction for interventions is provided by interventionists, classroom teachers, special education teachers, resource teachers, push-in teachers, or other staff, as determined by the RtI team.<br>• All interventions are documented.<br>• All instruction is monitored for fidelity of implementation. | | | |

*continued*

**FIGURE 7.12** *(continued)*

**Paradise View High School Shared Vision**

| **Needs Assessment** | **Purpose, Participants, and Outcomes** | | **Implementation** | | **Results** | |
|---|---|---|---|---|---|---|
| *What are the data showing about the need for the program or process?* | *What is the purpose of the program or process?* | *What are the intended outcomes?* | *How should the program/process be implemented with integrity and fidelity to ensure attainment of intended outcomes?* | *How is implementation being monitored?* | *How will results be measured?* | *What are the results?* |
| • Proficiency on the State Standards Assessment is as follows:<br>— ELA: 48% proficient<br>— Math: 16% proficient<br>• Attitudes: Not all students feel they are listened to by their teachers.<br>— Students do not feel challenged by the work they are asked to do.<br>— Students feel other students' behavior gets in the way of their learning.<br>— Teachers feel student behavior needs to improve. | | | • The RtI system is evaluated for implementation fidelity and effectiveness.<br>(See Figure 7.13 for the implementation flowchart.) | | | |

| ***Needs Assessment*** | ***Purpose, Participants, and Outcomes*** | | ***Implementation*** | | ***Results*** | |
|---|---|---|---|---|---|---|
| *What are the data showing about the need for the program or process?* | *What is the purpose of the program or process?* | *What are the intended outcomes?* | *How should the program/process be implemented with integrity and fidelity to ensure attainment of intended outcomes?* | *How is implementation being monitored?* | *How will results be measured?* | *What are the results?* |
| — Teachers do not work together to create a continuum of learning.<br>— Not all teachers believe that using data will improve student learning.<br>— Teachers say they believe all students can learn. | | | | | | |

FIGURE 7.13

**Paradise View High School Shared Vision Flowchart**

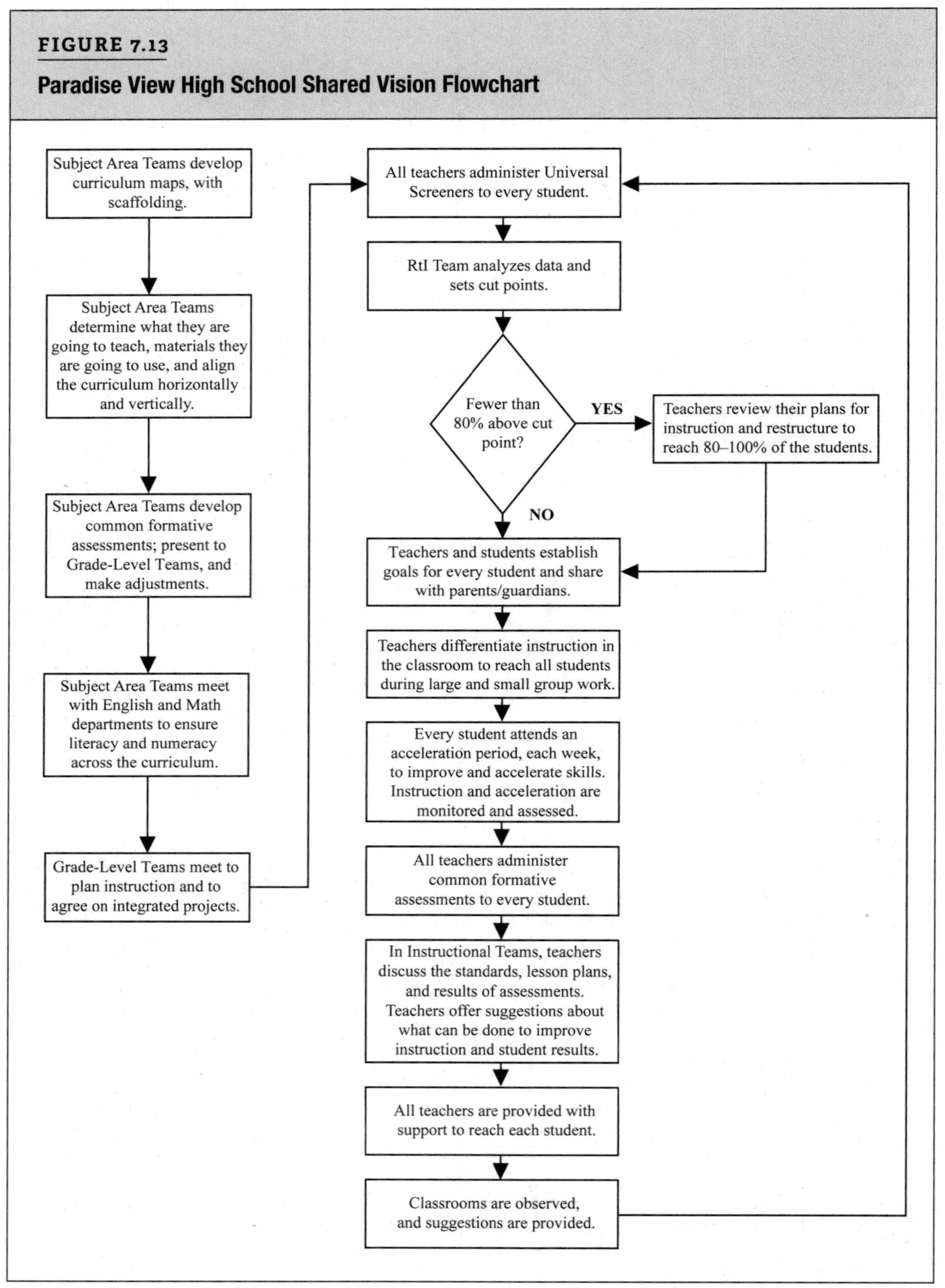

**FIGURE 7.14**

**Paradise View High School Logic Model**

## Context/Inputs

*School*
- Accountability ratings
- Attendance
- Culture/environment/safety
- Dropouts
- Enrollment
- Grade-level expectations
- Graduation rates
- Mobility rates
- Remediation
- Resources
- Special Education

*Students*
- Advanced Placement
- Attendance
- Attitudes
- Behavior/discipline
- Credits
- Dropout
- Extracurricular activities
- Gender, ethnicity/race, qualifiers for free/reduced-cost lunch, homeless, first language, second language, special education, gifted
- Health
- Honors
- Learning styles
- Middle school successes
- Mobility
- Pre-K attendance
- Retentions
- Tardies

*Teachers*
- Attendance
- Attitudes
- Gender, ethnicity/race
- Number of years teaching/number of years at this school
- Qualifications, training, experience
- Teaching style
- Core values and beliefs

## Processes

*Processes*
- Acceleration processes
- Alignment of curriculum, instruction, assessment
- Assessment (common, formative, progress monitoring, grades, standards)
- Behavior processes
- Business partnerships
- Differentiated instruction
- Grading policies
- Guidance process
- Homework
- Literacy across the curriculum
- Parent/family involvement
- Professional learning communities
- Professional learning for teachers, support staff, and principals
- Program evaluation
- Promotion/retention policies
- Response to Interventions (RtI)
- Standards curriculum
- Teacher collaboration
- Teaching strategies

*Programs*
- Advanced Placement
- After-school programs
- At-risk programs
- Athletic programs
- Career planning
- Career/technology education
- Debate and speech competitions
- English as a second language
- Guidance program
- Homeless program
- Honors
- Interventions
- Language programs
- Leadership
- Music program
- Peer Mediation
- Special Education Programs

## Outcomes

### *Short-Term*

*Students*
- Ability to plan for the future
- Able to compute
- Able to read
- Able to think critically
- Able to think scientifically
- Able to use technology effectively
- Able to write effectively
- Appreciation of the arts and music
- Excellent senior projects
- Feeling that teacher cares about them
- Feeling that they belong
- Good attendance
- High school graduation
- Involved in extracurricular activities
- Involved in school activities
- Knowledgeable of history
- Leadership
- Listening
- Logical reasoning
- Love school
- No repeat or make-up courses in high school
- Positive attitudes about learning
- Positive behavior
- Good grades
- Positive self-esteem
- Problem solving
- Speaking
- Student learning growth every year
- Successful SAT/ACT/State assessment results

*Administration and Staff*
- Caring relations with students
- Collaboration with one another
- Effective classrooms
- Engaged in partnerships with parents, business, and community
- Good morale
- Implementation of vision with integrity and fidelity
- Increased instructional skills

### *Long-Term*

*Students*
- College graduation—2-year, 4-year
- Community involvement
- Employment
- Enrollment in college—2-year, 4-year
- No remedial courses in college
- Scholarships
- Successful in college
- Successful in life
- Successful in work

*Parents*
- Able to support their childrens' lifelong ambitions

*Community*
- Benefiting from having great schools and employees

**FIGURE 7.14** *(continued)*

**Paradise View High School Logic Model**

| Context/Inputs | Processes | Outcomes: *Short-Term* | Outcomes: *Intermediate* |
|---|---|---|---|
| *Principals*<br>• Attendance<br>• Attitudes<br>• Gender, ethnicity/race<br>• Leadership style<br>• Number of years as administrator/principal<br>• Qualifications, training, experience<br>• Core values and beliefs | *Programs* continued<br>• Student teamwork<br>• Study skills program | *Administration and Staff* continued<br>• Increased teacher confidence in differentiating instruction and grouping students<br>• Program alignment to college and workforce<br>• Providing a physically and emotionally safe environment<br>• Satisfied with services they are offering<br>• Understanding the impact of their actions/policies on students<br>• Using data to guide decision making | |
| *Parents*<br>• Attitudes<br>• Educational background<br>• Language spoken at home<br>• Number of children in the home<br>• Number of parents/people in the home<br>• Participation in school events<br>• Socioeconomic status | | *Parents/Families*<br>• Engaged in partnership with teachers, principals, and students<br>• Involved with child's learning<br>• Knowledgeable of strategies to enhance childrens' learning<br>• Satisfied with services the school provides | |

more. That way, teaching would be easier for them. The principal knew the issue would be coming up so he had prepared data related to the matter. Collectively, staff reviewed the results for the students in question, and the math department, specifically. They could see that the new approach to RtI was helping the students. There was a discussion about other ways to improve their RtI process, and not have it be so labor intensive. Those changes were incorporated. Bottom line: the principal and staff agreed to continue with their RtI process, with enhancements. The principal told all staff, "You can implement it by yourself, or with help. Let the RtI team know if you need help. Not doing RtI, as designed, is not an option." The principal referred all staff to the expectations spelled out in the mission, vision, and the Program Evaluation Tools for the shared vision, programs, and processes, and RtI.

At the end of that school year, staff saw incredible improvements across the board on all their outcomes. They attributed these improvements to the intensive analysis of their data and processes, clarity of processes, implementation monitoring, and evaluation of the parts and the whole. By the end of the third year, the school's results were considered exemplary.

## Conclusion

This school had gone from failing to becoming a true learning organization through intensive planning, collaboration, beliefs in all students learning, implementation with fidelity, and comprehensive program and process evaluation.

# References and Resources


Anderson, B., Fagerhaug, T., Henriksen, B., & Onsoyen, L. (2008). *Mapping work processes.* (2nd ed.). Milwaukee, WI: ASQ Quality Press.

Balzer, W.K. (2010). *Learn higher education: Increasing the value and performance of university processes.* New York, NY: Productivity Press.

Barrett, N.F. (2013). *Program evaluation: A step-by-step guide* [Kindle Version]. Retrieved from Amazon.com

Bernhardt, V.L. & Hébert, C.L. (2017). *Response to intervention (RtI) and continuous school improvement (CSI): How to design, implement, monitor, and evaluate a schoolwide prevention system* (2nd ed.). New York, NY: Routledge.

Bernhardt, V.L. (2016). *Data, data everywhere: Bringing all the data together for continuous school improvement* (2nd ed.). New York, NY: Routledge.

Bernhardt, V.L. (2013). *Data analysis for continuous school improvement.* (3rd ed.). New York, NY: Routledge.

Bernhardt, V.L., & Geise, B.J. (2009). *From questions to actions: Using questionnaire data for continuous school improvement.* New York, NY: Routledge.

Black, P.J., & Wiliam, D. (1998). Inside the black box: Raising standards through classroom assessment. *Phi Delta Kappan, 80(2),* 139–148. Available from: *http://www.pdkintl.org/publications (archive)*

Damelio, R. (1996). *The basics of process mapping.* New York, NY: Productivity Press.

Davidson. E.J., (2013). *Actionable evaluation basics: Getting succinct answers to the most important questions* [Kindle Version]. Retrieved from Amazon.com

Earl, S., Carden, F, & Smutylo, T. (2001). *Outcome mapping: Building learning and reflection into development programs.* Ottawa, Ontario: IDRC Books.

Eisner, E. (2001). *The educational imagination: On the design and evaluation of school programs* (3rd ed.). Upper Saddle River, NJ: Prentice Hall.

Fitzpatrick, J., Sanders, J., & Worthen, B. (2004). *Program evaluation: Alternative approaches and practical guidelines* (3rd ed.). Boston, MA: Allyn & Bacon.

Fitzpatrick, J.L. & Morris, M. (1999). Current and emerging ethical challenges in evaluation. *New directions for evaluation, No. 82.* San Francisco, CA: Jossey-Bass.

Frechtling, J.A. (2007). *Logic modeling methods in program evaluation (research methods for the social sciences).* San Francisco, CA: Jossey-Bass.

Gudda, P. (2010). *A guide to project monitoring and evaluation* [Kindle Version]. Retrieved from Amazon.com

Jason, M.H. (2008). *Evaluation programs to increase student achievement.* (2nd ed.). Thousand Oaks, CA: Corwin Press, Inc.

Kaufman, R., Guerra, I., & Platt, W.A. (2006). *Practical evaluation for educators: Finding what works and what doesn't.* Thousand Oaks, CA: Corwin Press, Inc.

Killion, J. (2008). *Assessing impact: Evaluating staff development* (2nd ed.). Thousand Oaks, CA: Corwin Press. A joint publication with the National Staff Development Council.

Knowlton, L.W., & Phillips, C.C. (2013). *The logic model guidebook: Better strategies for great results* (2nd ed.). Thousand Oaks, CA: Sage Publications, Inc.

Leeuw, F.L. (2003). Reconstructing program theories: Methods available and problems to be solved. *American Journal of Evaluation*, Vol. 24, No. 1, pp. 5–20. Thousand Oaks, CA: Sage Publications, Inc. Available from: *http://aje.sagepub.com/cgi/content/abstract/24/1/5*

Madison, D.J. (2005). *Process mapping, process improvement, and process management: A practical guide to enhancing work and information flow.* Chico, CA: Paton Press, LLC.

Markiweicz, A. & Patrick, I. (2016). *Developing monitoring and evaluation frameworks.* Thousand Oaks, CA: Sage Publications, Inc.

McDavid, J.C. & Hawthorn, L.R.L. (2006). *Program evaluation and performance measurement: An introduction to practice.* Thousand Oaks, CA: Sage Publications, Inc.

McKinsey & Company (April 2009). *Detailed findings on the economic impact of the achievement gap in America's schools.* Available from: *http://www.mckinsey.com/clientservice/socialsector/detailed_achievement_gap_findings.pdf*

Newmann, F., Smith, B., Allensworth, E., & Bryk, A. (2001). Instructional program coherence: What it is and why it should guide school improvement policy. *Educational Evaluation and Policy Analysis, 23*(4), 297–321.

Oriel Incorporated (2002). *Flowcharts: Plain & simple.* Madison, WI.

Parsons, Beverly A. (2002). *Evaluative inquiry: Using evaluation to promote student success.* Thousand Oaks, CA: Corwin Press, Inc.

Patton, M.Q. (2012). *Essentials of utilization-focused evaluation.* Thousand Oaks, CA: Sage Publications, Inc.

Patton, M.Q. (2011). *Developmental evaluation: Applying complexity concepts to enhance innovation and use.* New York, NY: the Guilford Press.

Patton, M.Q. (2001). *Qualitative research and evaluation methods* (3rd ed.). Newbury Park, CA: Sage Publications, Inc.

Ridge, J.B. (2010). *Evaluation techniques for difficult-to-measure programs: For education, nonprofit, grant-funded, business and human service programs* [Kindle Version]. Retrieved from Amazon .com

Rogers, P.J., Hacsi, T.A., Petrosino, A., & Huebner, T.A. (2000). Program theory in evaluation challenges and opportunities. *New directions for evaluation, No. 87.* San Francisco, CA: Jossey-Bass.

Rossi, P.H., Freeman, H.E., & Lipsey, M.W. (2003). *Evaluation: A systematic approach* (7th ed.). Thousand Oaks, CA: Sage Publications, Inc.

Sanders, J.R., & Sullins, C.D. (2006). *Evaluating school programs: An educator's guide* (3rd ed.). Thousand Oaks, CA: Corwin Press, Inc.

Saunders, R.P, (2016). *Implementation monitoring and process evaluation.* Thousand Oaks, CA: Sage Publications, Inc.

Senge, P.M. (2006). *The fifth discipline: The art and practice of the learning organization.* New York, NY: Currency Doubleday.

Singapore American School. (August, 2016). Retrieved from sas.edu.sg

Stufflebeam, D.L. & Coryn, C.L.S. (2014). *Evaluation theory, models, and applications* (Research methods for the social sciences). San Francisco, CA: Jossey-Bass.

Tague, N.R. (2005). *The quality toolbox* (2nd ed.). Milwaukee, WI: ASQ Quality Press.

University of Wisconsin Cooperative Extension, *Program Development and Evaluation.* Available from: *http://www.uwex.edu/ces/pdande/index.html*

W. K. Kellogg Foundation (2006). *W. K. Kellogg Foundation logic model development guide: Using logic models to bring together planning, evaluation, and action.* Battle Creek, MI. Available from: *www.wkkf.org*

# Index

The letter *f* following a page number denotes a figure.

# About the Author

Victoria L. Bernhardt, PhD, is Executive Director of the Education for the Future Initiative, whose mission is to build the capacity of learning organizations at all levels to gather, analyze, and use data to continuously improve learning for all students. She is also a professor emeritus in the College of Communication and Education, at California State University, Chico.

Dr. Bernhardt received her PhD in Educational Psychology Research and Measurement, with a minor in Mathematics, from the University of Oregon. She has worked for more than 25 years with learning organizations all over the world to assist them with their continuous improvement and comprehensive data analysis work. She is the author of 20 books, including the widely recognized *Data Analysis for Continuous School Improvement, 3rd Edition*. Victoria is passionate about her mission of helping educators continuously improve teaching and learning by gathering, analyzing, and using data. She provides consultations, professional learning, and keynote addresses for schools and the agencies that serve them on the topics of comprehensive data analysis, Response to Intervention, and continuous school improvement. To learn more about Victoria's work, visit http://eff.csuchico.edu.

## Related Resources

At the time of publication, the following ASCD resources were available (ASCD stock numbers appear in parentheses). For up-to-date information about ASCD resources, go to www.ascd.org. You can search the complete archives of Educational Leadership at http://www.ascd.org/el.

**ASCD EDge® Group**

Exchange ideas and connect with other educators interested in change and school culture on the social networking site ASCD EDge at http://ascdedge.ascd.org/.

**Print Products**

*Educational Leadership: Resilience and Learning* (September 2013) (#114018)

*Inclusive Schools in Action: Making Differences Ordinary* by James McLeskey and Nancy L. Waldron (#100210)

*Teaching in Tandem: Effective Co-Teaching in the Inclusive Classroom* by Gloria Lodato Wilson and Joan Blednick (#110029)

*Educational Leadership: Doing Data Right* (November 2015) (#116030)

*Linking Teacher Evaluation and Student Learning* by Pamela D. Tucker and James H. Stronge (#104136)

For more information, send e-mail to member@ascd.org; call 1-800-933-2723 or 703-578-9600, press 2; send a fax to 703-575-5400; or write to Information Services, ASCD, 1703 N. Beauregard St., Alexandria, VA 22311-1714 USA.